Constructing a Microprogrammed Computer

```
;======================================================
            ; Fetch machine instruction, increment pc
fetch       a.pc c.mar
            a.pc add b.1 c.pc rd


;======================================================
            ; Decode instruction

            a.mdr add b.0 c.ir neg@L1
L0          a.ir sll b.1 c.dc neg@L01
L00         a.dc sll b.1 c.dc neg@L001
L000        a.dc sll b.1 c.dc neg@L0001
            br@L0000
```

Anthony J. Dos Reis

Constructing a Microprogrammed Computer edition 1
Copyright © 2019 by Anthony J. Dos Reis, all rights reserved

ISBN: 9781072708995

Preface

Constructing a Microprogrammed Computer (*CMC*) along with its companion volume, *C and C++ Under the Hood* (*CUH*), will provide you with a detailed examination of a complete computer system. *CMC* covers the hardware and firmware side of the system (digital circuits and microprogramming); *CUH* covers the software side of the system (assemblers, compilers, linkers, C, and C++). Neither book is the prerequisite of the other. If you prefer the top-down approach, read *CUH* first. If you prefer the bottom-up approach, read *CMC* first.

One distinguishing characteristic of both *CMC* and *CUH* is that they give you plenty to do. Thus, you will learn not only by reading but also by doing. For example, *CMC* will guide you through the microcode implementation of three machine instruction sets: a basic register instruction set, a stack instruction set, and an "optimal" register instruction set (the optimal instruction set is used throughout *CUH*). *CUH* will guide you through the implementation of a machine interpreter, an assembler, a linker, and several other system programming projects.

To get the latest version of the software package, send an email to cmctextbook@gmail.com. You will then immediately receive an automatic reply with a link to the site at which you can download the software package (cmc.zip). The software package run on Windows, Mac OS X, Linux, and Raspbian.

Anthony J. Dos Reis
SUNY New Paltz

Table of Contents

Preface, *iii*

1 Numbering Systems

Decimal, Binary, Hexadecimal, 1
Numbering Bits, 2
Adding Positional Numbers, 3
Representing Negative Numbers, 4
Signed and Unsigned Numbers, 5
Range of Signed and Unsigned Numbers, 6
Converting between Binary and Hex, 7
Converting Decimal to Binary, 8
Zero and Sign Extension, 9
Problems, 9

2 Machine Language

Introduction, 11
Structure of the LCC, 11
Simple Machine Language Program, 13
Load Immediate Instruction, 17
Strings, 17
Branch Instructions, 19
Trap Instructions, 20
Problems, 22

3 Assembly Language

Introduction, 23
Load Immediate Instruction, 26
.Blkw and .Stringz Directives, 27
Branch Instructions and Loops, 28
Accessing the Stack, 29
Calling Subroutines, 31
Passing Arguments Via the Stack, 33
Basic Instruction Set Summary, 34
Problems, 36

4 Simple Digital Circuits

Introduction, 38
Transistors, 38
NOT Gate, 40
XOR, OR, NOR, AND, and NAND Gates, 41
Tri-State Buffer, 43

Realizing Boolean Functions with Digital Logic, 45
   Half and Full Adders, 46
   Sequential Circuits, 48
   Problems, 50

5 Complex Digital Circuits

   Introduction, 53
   16-bit AND, OR, XOR, and NOT Circuits, 53
   SEXT Circuit, 54
   Multiplexer, 54
   Decoder, 55
   Multi-Bit Addition and Subtraction, 56
   Signed Overflow, 57
   Unsigned Overflow, 59
   Barrel Shifter, 61
   Multiplier and Divider/Remainder Circuits, 63
   Random-Access and Read-Only Memory, 63
   Registers, 65
   Arithmetic-Logic Unit, 65
   Loadable Binary Counter, 67
   Clock Sequencer, 68
   Problems, 70

6 Microlevel of the LCC

   Introduction, 72
   Data Path, 72
   Main Memory Interface, 73
   Decoding the Registers Fields in a Microinstruction, 73
   Specifying the ALU Operation, 75
   Branch-Control Logic, 76
   Complete Microlevel of the LCC, 77
   Problems, 78

7 Microprogramming the Basic Instruction Set

   Introduction, 79
   Format of a Microinstruction, 79
   Alternate Register Names, 79
   Symbolic Microcode, 81
   Fetching a Machine Instruction and Decoding the Opcode, 83
   Interpreting Machine Instructions with Microcode, 86
   Assembling and Using Microcode, 87
   Debugging Microcode, 89
   Compiling C Code to the Basic Instruction Set, 90
   Problems, 93

8 Microprogramming the Stack Instruction Set

Pointers, 96
Non-Constant Relative Addresses, 98
Stack Architecture, 99
Frame Pointer Register, 100
Dereferencing Pointers, 105
Multiplying, 106
Adding Opcodes, 108
Stack Instruction Set Summary, 109
Problems, 111

9 Microprogramming Optimal Instruction Set

Flaws in the Stack Instruction Set, 115
Instruction Set Architecture, 116
Programming with the Optimal Instruction Set, 123
Amux, Bmux, and Cmux Multiplexers, 125
Complications with the St Instruction, 128
User and System Flag Registers, 128
Epilog, 130
Problems, 130

Appendix A: ASCII, 133
Appendix B: Basic Instruction Set Summary, 134
Appendix C: Stack Instruction Set Summary, 135
Appendix D: Optimal Instruction Set Summary, 136
Appendix E: Microinstruction Format, 137
Appendix F: Default Register Names, 139
Index, 140

1 Numbering Systems

Decimal, Binary, and Hexadecimal

Decimal is a positional numbering system. It is so called because in a decimal number the contribution of each digit to the value of the number depends not only on the digit but on its position in the number. For example, consider the three-digit decimal number 123:

$$\frac{1 \quad 2 \quad 3}{100 \quad 10 \quad 1} \quad \text{weights}$$

Each position has a weight. In a whole number, weights start with 1 and increase from right to left by a factor of 10 from each position to the next. The value of the number is given by the sum of each digit times its weight. Thus, the value of 123 is

$1 \times 100 + 2 \times 10 + 3 \times 1$

The 1 digit contributes $1 \times 100 = 100$ to the value of the number; the 2 digit contributes $2 \times 10 = 20$ to the value of the number, the 3 digit contributes $3 \times 1 = 3$ to the value of the number. Although the 3 digit is greater than the 2 digit, the 2 digit contributes more to the value of the number than the 3 digit because its weight is ten times that of the 3 digit.

We call decimal the *base-10* numbering system because it uses 10 distinct symbols and because weights increase by a factor of 10 from each position to the next. *Binary* is the *base-2* numbering system. It uses two distinct symbols (0 and 1), called *bits*. In binary, position weights increase by a factor of 2 from each position to the next. For example, consider the five-bit binary number 01101:

$$\frac{0 \quad 1 \quad 1 \quad 0 \quad 1}{16 \quad 8 \quad 4 \quad 2 \quad 1} \quad \text{weights}$$

Its value is given by the sum of each digit times its weight:

$0 \times 16 + 1 \times 8 + 1 \times 4 + 0 \times 2 + 1 \times 1 = 13$ decimal

It is easy to determine the decimal value of a binary number: Simply add up the weights corresponding to the 1 bits. In the binary number above, the weights corresponding to 1 bits are 8, 4, and 1. Thus, the value of the number is $8 + 4 + 1 = 13$ decimal.

We call a sequence of eight bits a *byte*. For example, 1111000010101010 consists of two bytes: 11110000 and 10101010. To *complement* a bit means to flip it. That is, change a 1 bit to 0, and a 0 bit to 1.

Hexadecimal (or hex for short) is the *base-16* numbering system. It uses 16 symbols: 0 to 9 and A, B, C, D, E, and F in upper or lower case (the lowercase forms are more convenient for keyboard input because their entry do not require the shift key). The values of A, B, C, D, E, and F and their corresponding lowercase forms equal decimal 10, 11, 12, 13, 14, and 15, respectively.

Weights in a hexadecimal number increase by a factor of 16. For example, consider the three-digit hex number 2C5:

$$\frac{2 \quad C \quad 5}{256 \quad 16 \quad 1} \text{ weights (in decimal)}$$

Its value is given by

$$2\times256 + C\times16 + 5\times1$$

The hex digit C is 12 in decimal so the expression above using only decimal is equal to

$$2\times256 + 12\times16 + 5\times1 = 512 + 192 + 5 = 709$$

The following table shows the decimal numbers from 0 to 15 and their binary and hex equivalents.

Decimal	Binary	Hex
0	0000	0
1	0001	1
2	0010	2
3	0011	3
4	0100	4
5	0101	5
6	0110	6
7	0111	7
8	1000	8
9	1001	9
10	1010	A (or a)
11	1011	B (or b)
12	1100	C (or c)
13	1101	D (or d)
14	1110	E (or e)
15	1111	F (or f)

Since we will be working quite a bit with binary and hex, *it is essential that you memorize this table.*

If you append a 0 on its right side of a binary whole number, the weight of each digit increases by a factor of 2. Thus, the value of the number doubles. For example, 3 in binary is 0011. If we append a 0, we get 0110, which is 6 decimal. If we append another 0, we get 1100, which is 12 decimal. Of course, adding a 0 on the left side does not affect the value of a number. For example, 0110 equals 00110.

Similarly, if you append a 0 on the right side of a hex whole number, its value increases by a factor of 16 (since the weight of each digit increases by a factor of 16). For example, A is 10 decimal, and A0 is 160 decimal.

Rule: Adding a 0 on the right side of a positional whole number multiplies its value by its base.

Numbering Bits

The bits in a binary number are numbered right to left starting with 0. For example, in an eight-bit number, the rightmost bit is bit 0; the leftmost bit is bit 7:

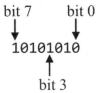

With this numbering scheme, there is a nice correspondence between a bit's number and its weight: Bit i has weight 2^i. For example, bit 3 in the binary number above has the weight $2^3 = 8$. Thus, it contributes 8 to the value of the number.

The leftmost bit and the rightmost bit in a binary number are called the *most significant bit* (abbreviated msb) and the *least significant bit* (abbreviated lsb), respectively.

Adding Positional Numbers

Let's quickly review how we add two decimal numbers. Consider the following addition:

```
    1   carries
  157
+ 238
  ---
  395
```

We start in the right column. Adding 7 and 8, we get 15. The result is two digits. So we record the right digit 5 at the bottom of the column and carry the left digit 1 to the next column. Thus, in the next column we add 1, 5, and 3 to get 9. The result is a single digit so we do not carry into the next column. Finally, we add 1 and 2 in the left column to get 3 for that column.

To add two binary numbers, we take exactly the same approach as we take with decimal. For example, consider the following addition of the binary numbers 0011 and 0011:

```
   11   carries
 0011
+0011
 ----
 0110
```

When we add the right column, we get 10 binary (2 decimal). The result is two bits. So we record the right bit 0 and carry the left bit 1 to the next column. Thus, in the next column, we add 1, 1, and 1 to get 11 binary (3 decimal). So we record the right bit 1 and carry the left bit 1 to the next column, where we add 1, 0, and 0 to get 1. Finally, in the leftmost column, we add 0 and 0 to get 0.

Let's now add the hex numbers 1B and 37:

```
   1   carries
  1B
+ 37
  --
  52
```

Adding B (11 in decimal) and 7, we get 12 hex (18 decimal). We record the 2 digit and carry 1 to the next column, where we add 1, 1, and 3 to get 5.

Representing Negative Numbers

Signed numbers within a computer are usually represented in the *two's complement* system. Before we discuss two's complement, let's make a simple observation. Suppose a computer represents numbers using only 4 bits. If our computer adds 0001 to 1111, what is the result? Here is the addition:

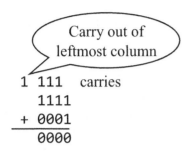

```
1 111    carries
  1111
+ 0001
  ----
  0000
```

We get zero with a carry out of the leftmost column. Since we are assuming the computer uses only four bits to represent numbers, this carry in not included in the result. Thus, the result is 0000.

Rule: Adding 1 to a binary number with a fixed number of bits all of which are 1 results in all zeros.

Let's now experimentally determine the two's complement representation of −3. We want the binary form of −3 that when added to the binary form of +3 gives a sum of zero. Let's see if complementing (i.e., flipping) all the bits in the binary form of +3 is the desired representation of −3:

```
  0011  = +3
+ 1100  = +3 with each bit flipped
  ----
  1111
```

We do not get zero so 1100 is not −3. But recall our preceding rule: Adding 1 to all 1's gives zero. Thus, because flipping the bits of +3 gives all 1's when added to +3, flipping the bits *and* adding 1 should give us the representation that produces zero when added to +3. Let's try it. Flipping the bits of +3 and adding 1, we get

```
  1100  = +3  with each bit flipped
+ 0001  add 1
  ----
  1101  Is this −3?
```

Is 1101 the representation of −3 that we want? Let's add it to +3 to see if it gives zero:

```
1 111    carries
  0011  = +3
+ 1101  Is this −3?
  ----
  0000
```

We, indeed, get zero with a carry out of the leftmost position. Thus, 1101 is the correct representation of −3 in the two's complement system.

Rule: To negate a binary number in the two's complement system, flip its bits and add 1.

The binary number system that can represent both positive and negative numbers and in which a number is negated by flipping its bits and adding 1 is called the *two's complement system*. We call the negation of a number the *two's complement* of that number. For example, the two's complement of 0011 (+3) is 1101 (−3). The two's complement of 1101 (−3) should get us back to is 0011 (+3). Indeed, it does:

```
  0010 = −3 with each bit flipped
+    1
  0011 = +3
```

To add two's complement numbers, we simply add them using the standard adding procedure. It does not matter if one is positive and one is negative. For example, lets add +1 and −3. The result should be the two's complement number for −2:

```
  0001 = +1
+ 1101 = −3
  1110 = −2
```

To confirm that 1110 is −2, take its two's complement to see if you get +2. That is, flip the bits in 1110 and add 1. The result is indeed 0010 (+2), which confirms that 1101 is −2.

In the two's complement system, −1 is represented with all 1's. Let's confirm this by taking the two's complement of +1. We flip the bits in +1 and add 1. We get

```
  1110 = +1 with each bit flipped
+    1
  1111 = −1
```

Rule: In the two's complement system, all 1's represents −1.

In the two's complement system, the leftmost bit of a number indicates the sign of the number: A 1 bit indicates the number is negative; a 0 bit indicates the number is non-negative (i.e., zero or positive). Note that in the two's complement system, the bits to the right of the sign bit do *not* represent the magnitude of the number. For example, 1111 in the two's complement system is −1. The three bits to the right of the sign bit, 111 (7 decimal), is *not* the magnitude of the number.

Signed and Unsigned Numbers

If the number representation used for a number allows for positive and negative numbers, we say the number is a *signed number*. Otherwise it is an *unsigned number*. Because unsigned numbers have no sign, they represent only non-negative numbers. For example, the value of the unsigned number 1111 is 15 in decimal, but as a two's complement signed number, its value is −1. Note that the number 1111 can be either an unsigned number or a signed number.

What makes a binary number signed or unsigned is how it is treated. For example, suppose you compare 1111 and 0010 and conclude that 1111 is bigger (because 1111 represents 15 and 0010 represents 2). Then the numbers are unsigned because you are treating them that way. But if you conclude 0010 is bigger (because 0010 represents 2 and 1111 represents −1), then the numbers are signed.

From this point on, when we use the term "signed number," we mean a number in the two's complement system.

Range of Signed and Unsigned Numbers

There are two patterns that can be represented by a single bit: either 0 or 1. With two bits, either bit can be 0 or 1. Thus, there are $2 \times 2 = 2^2 = 4$ patterns: 00, 01, 10, 11. With three bits, there are $2 \times 2 \times 2 = 2^3 = 8$ patterns: 000, 001, 010, 011, 100, 101, 110, 111. Generalizing, with n bits we get 2^n patterns.

If we represent unsigned numbers with four bits, we can have $2^4 = 16$ patterns. If we use these 2^4 patterns to represent the sequence of non-negative numbers starting with 0, we can represent the numbers 0 to $2^4 - 1$ (we go up to $2^4 - 1 = 23$, not 2^4, because we are starting from 0). Generalizing, with n bits we can represent unsigned numbers from 0 to $2^n - 1$. For example, with eight bits, we can represent unsigned numbers from 0 to $2^8 - 1 = 255$. The following table shows the range of four-bit signed and unsigned numbers:

Unsigned	Value	Signed	Value
0000	0	1000	-8
0001	1	1001	-7
0010	2	1010	-6
0011	3	1011	-5
0100	4	1100	-4
0101	5	1101	-3
0110	6	1110	-2
0111	7	1111	-1
1000	8	0000	0
1001	9	0001	1
1010	10	0010	2
1011	11	0011	3
1100	12	0100	4
1101	13	0101	5
1110	14	0110	6
1111	15	0111	7

With two's complement signed numbers, the left bit indicates the sign (0 for non-negative numbers or 1 or negative numbers). Suppose we represent numbers with four bits. For the negative numbers, the sign bit is 1, leaving only three bits to specify the negative number. With three bits, we can specify $2^3 = 8$ numbers. Thus, starting from -1, we can represent the numbers -1 down to -8. For the non-negative numbers, the sign bit is 0, leaving only three bits to specify the number. Thus, as with the negative numbers, we can represent $2^3 = 8$ non-negative numbers. But we start from 0, not 1. Thus, we can represent the numbers 0 to 7 (not 1 to 8). The table above shows the four-bit unsigned and signed numbers in ascending order along with their values in decimal. Note for the signed numbers, the negative numbers go down to -8, but the non-negative number go up to only +7. The numbers go one further in the negative direction than in the positive direction because the negative numbers start from -1, but the non-negative numbers start from 0.

The following table shows the ranges of unsigned and signed numbers with n bits for several values of n.

n	Unsigned	Signed
1	0 to 1	-1 to 0
2	0 to 3	-2 to 1
3	0 to 7	-4 to 3
4	0 to 15	-8 to 7
5	0 to 31	-16 to 15
6	0 to 63	-32 to 31
7	0 to 127	-64 to 63
8	0 to 255	-128 to 127
9	0 to 511	-256 to 255
10	0 to 1023	-512 to 511
12	0 to 4095	-2048 to 2047
16	0 to 65535	-32768 to 32767
k	0 to 2^k-1	-2^{k-1} to $2^{k-1}-1$

Do not attempt to memorize this table. Instead, learn the value of 2^n for n from 1 to 16. Once you know these powers of 2, it is easy to figure out the ranges of unsigned and signed numbers with n bits for the values of n in the table. For example, 8 bits has 256 patterns (because $2^8 = 256$). Thus, 8-bit unsigned numbers range from 0 to $2^8 - 1 = 255$. For signed numbers, half of the 2^8 patterns are for negative numbers, and half are for non-negative numbers. Half of 2^8 is $2^7 = 128$. Thus, 8-bit signed numbers range from -128 to 127. Here are the powers of 2 you should know:

n	2^n
1	2
2	4
3	8
4	16
5	32
6	64
7	128
8	256
9	512
10	1,024 (aka 1K)
11	2,048 (aka 2K)
12	4,096 (aka 4K)
15	32,768 (aka 32K)
16	65,536 (aka 64K)
20	1,048,576 (aka 1M)
30	1,073,741,824 (aka 1G)

Converting Between Binary and Hex

It is trivial to convert between binary and hex once you know the binary numbers from 0000 to 1111 and their hex equivalents. To convert a binary number to hex, break up the binary number into groups of four bits, starting from its right end. Then substitute the hex equivalent for each four-bit group. For example, to convert 110101110000001100, we first break it up into four-bit groups:

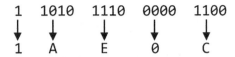

We then substitute the hex equivalent for each four-bit group:

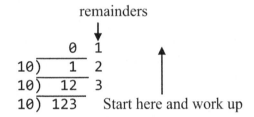

Thus, 11010111000001100 binary is equal to 1AE0C hex. To convert hex to binary, we simply substitute the four-bit binary equivalent for each hex digit. For example, to convert A5 to binary, substitute 1010 for A and 0101 for 5 to get 10100101.

Converting Decimal to Binary

If we repeatedly divide a number by 10 until we get a 0 quotient, the remainders will be the digits that represent that number in decimal. For example, let's divide 123 by 10 repeatedly:

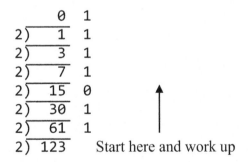

As you can see, the remainders are the digits that make up the decimal number. If, instead, we repeatedly divide by 2, then the remainders will be the bits that make up the binary number equal to 123 decimal:

```
        0  1
    2)  1  1
    2)  3  1
    2)  7  1
    2) 15  0
    2) 30  1
    2) 61  1
    2) 123     Start here and work up
```

Reading the remainders from the top down, we get the bits that make up the binary number equal to 123 decimal. Thus, 123 decimal = 1111011 binary. Let's check our answer by converting it to hex and then to decimal. 1111011 = 111 1011 = 7B hex = 7×16 + 11 = 123 decimal. You can similarly convert numbers to hex by dividing repeatedly by 16.

Zero and Sign Extension

Suppose we have a one-byte binary number that we want to extend to two bytes (recall that a byte is eight bits). We can do this in two ways. We can add eight zeros on the left, or add eight copies of the sign bit on the left. For example, to extend 11111111, we can add eight zeros to get

    0000000011111111

or we can replicate the sign bit (i.e., its leftmost bit) of 11111111 to get

    1111111111111111

The former approach is called *zero extension*; the latter approach, *sign extension*.

If a negative signed number is zero-extended, it changes its value. 11111111 (which is equal to -1) zero-extended to 16 bits is 0000000011111111 (which is equal to +255). If, however, it is sign-extended, its value remains -1.

Rule: Always sign-extend signed numbers.

If an unsigned number with 1 in its leftmost position is sign-extended, its value changes. For example, if the unsigned number 11111111 (255) is sign-extended to 16 bits, we get 1111111111111111 (65535). However, if we zero-extend an unsigned number, its value remains the same.

Rule: Always zero-extend unsigned numbers.

Problems

1) Convert the following binary numbers to decimal:

    0101111101101111, 11111111, 1000000000

2) Convert the binary numbers in the preceding problem to hex.

3) Add 8000 hex (-32768 decimal) and ffff hex (-1 decimal). Represent the computed result using 16 bits. What is the sign of the computed result? What is the sign of the true result? Why is there a discrepancy between the computed result and the true result?

4) What is the range of 12-bit unsigned numbers and 12-bit two's complement signed numbers?

5) What is the range of 5-bit two's complement numbers? 9-bit? 11-bit?

6) Convert the following decimal numbers to binary and hexadecimal:

    1023, 1024, 1025, 255, 16

7) Convert the following hexadecimal numbers to binary:

   5567 ABABAB, F03, 3579BDF, 2468ACE, FCC

8) Convert the following hexadecimal numbers to decimal:

   A0, B0, C0, D0, E0, F0, 400, 10000

9) Add the following pairs of binary numbers: Give your answers in both binary and hex.

   ```
   0111111111111111      0111000111000111      0011111111111111
   0000101010101011      0010101010101010      0000000000000001
   ```

10) Subtract the numbers in the preceding question.

11) Add the following pairs of hexadecimal numbers:

    ```
    0FFFFFFF          996
    000000001         959
    ```

12) Subtract the numbers in problem 11. Give your answers in both hex and decimal.

13) Write −75 decimal as a 16-bit two's complement binary number.

14) What is the next (and final number) in this sequence: 1000, 22, 20, 13, 12, 11, 10?

15) Convert 0.111 binary to decimal.

16) Convert 0.5 decimal to binary.

17) Convert 0.75 decimal to binary.

18) Convert 0.1 decimal to binary. *Hint*: Multiply repeatedly by 2, removing whole part after each multiplication. The whole parts make up the binary number.

19) Convert the following octal (base 8) numbers to decimal: 123, 777, 100.

20) Convert the following base 9 numbers to octal (base 8): 123, 777, 100.

2 Machine Language

Introduction

Machine language is the only language the computer hardware can "understand." Thus, if you write a program in any language other than machine language, it first has to be translated to machine language before it can be executed by the computer. A machine language instruction is a binary number. A machine language program consists of a sequence of binary numbers.

Each type of computer has its own machine language. IBM mainframe computers have one type of machine language. PCs that run Windows have another type of machine language. In this book, we will study the LCC (**L**ow **C**ost **C**omputer). The LCC is a modification and extension of the computer model presented in Patt and Patel's book, *Introduction to Computing Systems*. It is a *microprogrammed computer*. That is, within the its central processing unit (CPU), there is a read-only memory, called *microstore*, that contains a *microprogram*. The microprogram consists of a sequence of microinstructions that determines the machine language of the LCC. Thus, by changing the microprogram, we can change the machine instructions that the LCC supports.

With high-level languages like C++, we often refer to the instructions that make up a program as "code." Similarly, we refer to the microinstructions that make up a microprogram as *microcode*. In subsequent chapters you will learn how to write the microcode that defines a machine language instruction set. By doing so, you will gain a clear understanding of the operation of the LCC at the microlevel. In addition, you will get a sense of what constitutes a good instruction set.

All the data, instructions, and addresses inside a computer are in binary. It is hard to read binary (for us—not for the computer), and binary numbers require a lot of space on the printed page. For this reason, in most of this book, we use hexadecimal to represent the binary numbers within the LCC.

Structure of the LCC

The two principal units of the LCC are the *central processing unit* (CPU) and *main memory* (see Fig. 2.1). Within the CPU are the *arithmetic/logic unit* (ALU) and the *control unit*. The ALU is the unit that performs high-speed computations. The control unit is the control center for all the other components of the computer. Within the control unit is the microstore that holds the microprogram that determines the machine instruction set. Also, within the CPU are storage areas, called *registers*, named `pc` (program counter), `ac` (accumulator), `sp` (stack pointer), and `ir` (instruction register). Each of these registers can hold one 16-bit number. The `pc` (program counter) "points to" the machine instruction in memory to be executed next. The `ac` register accumulates the results of computations. The `sp` register points to the top of the stack that resides in memory. The `ir` holds the machine instruction that the CPU is executing. We will elaborate on the function of each of these registers shortly

Memory is an array of $2^{16} = 65536$ cells, each of which can hold one 16-bit number. The memory cells are numbered starting with 0. The number of a cell (i.e., the number that identifies a particular cell) is called the *address* of that cell. The number *inside* the memory cell is the *contents* of that cell.

The *word size* of the LCC is 16 bits. That is, the computational circuits in the CPU of the LCC operate on units of data that are 16 bits wide. For example, the adder circuit in the CPU can add two 16-bit numbers.

In the LCC, each memory cell can hold one word (i.e., 16 bits). For this reason, we say its memory is *word addressable*. That is, successive addresses correspond to successive words. Memory on most computers, however, is *byte addressable*. That is, successive addresses correspond to successive bytes.

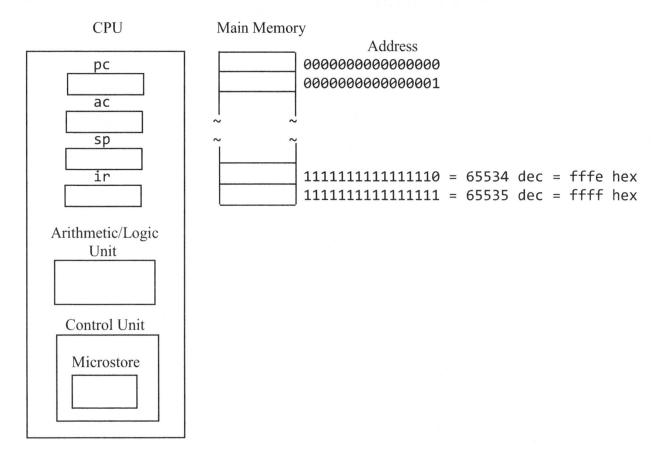

Figure 2.1

To execute a machine language program, the computer user enters a command to the operating system (OS) specifying the name of the file that holds the machine language program. The OS responds by loading the program into memory starting at some address, referred to as the *load point*. To keep our discussion as simple as possible in this introduction, let's assume the load point is always 0, and execution always starts at that address.

After the OS loads the program into memory, it loads the pc register with the load point (we are assuming it is the address 0). The CPU then executes the loop in Fig. 2.2 (a *loop* is a sequence of operations that are repeatedly executed). The first time the loop is executed, the pc register contains 0. Thus, in step 1, the CPU fetches the instruction from the memory cell at the address 0 and loads it into the ir. In step 2, it increments the pc register to 1. In step 3, it *decodes* the instruction (i.e., determine its opcode). In step 4, it executes the instruction it fetched in step 1, which is now in the ir. On the next iteration of the loop, the CPU fetches the instruction at the address 1—not 0—because the pc register now contains 1 (because of step 2 in the preceding iteration). Each time the CPU performs step 1, it gets the next instruction from memory because step 2 in the preceding iteration increments the pc register. Thus, the CPU executes instructions one after another in order of memory address. This process continues until a halt or branch machine instruction is executed. A halt machine instruction halts the execution of the program and causes a return to the OS. The branch machine instructions do not halt execution. Instead, they cause the CPU to go to a new location and start executing instructions from there. For example, a branch instruction at address 10 can cause the CPU to go back to address 0 and execute instructions in memory order starting again from there.

1. *Fetch* the instruction the `pc` register "points to." That is, the CPU loads the `ir` with the instruction in the memory cell whose address is in the `pc` register. The CPU does not remove the instruction from its memory cell. Instead, it makes a copy of it. Thus, the contents of the memory cell that the `pc` register points to are unaffected.

2. *Increment* the `pc` register.

3. *Decode* instruction.

4. *Execute* the instruction in the `ir`.

Figure 2.2

Simple Machine Language Program

As we mentioned above, the microprogram within the CPU determines the set of machine instructions the computer can execute. In this chapter, we will use the LCC with a microprogram that defines an instruction set we call the *basic instruction set*. It is an instruction set that is easy to understand and use. Thus, it is a good instruction set to use as we start to investigate the operation of the LCC.

Let's examine a simple machine language program that uses the basic instruction set. It consists of six instructions, each occupying one word in memory, and three data words. Let's assume this program is loaded into memory starting at the address 0. Here is a description of each word of the program along with its address:

Address (hex)	Description of Instruction
0:	Load a copy of the number in memory at address `0006` into the `ac` register.
1:	Add a copy of the number in memory at address `0007` to the `ac` register.
2:	Store the number in the `ac` register into the memory location at address `0008`.
3:	Display in decimal the number in the `ac` register.
4:	Move the display cursor to the beginning of the next line on the screen.
5:	Halt.
6:	First number (`0002` hex)
7:	Second number (`0003` hex)
8:	Location into which the sum is stored

The instruction at address 0 is a `ld` (load) instruction. Here is the instruction *in binary*:

ld opcode   12-bit memory address (006 hex)
0000   000000000110

The first four bits (0000) is the *opcode*. The opcode specifies the operation to be performed. 0000 is the opcode for the ld instruction. The next twelve bits—000000000110—zero-extended to 16 bits is the memory address of the number to be loaded into the ac register. When this ld instruction is executed, the CPU loads a copy of the number at this address into the ac register, overlaying whatever is there. The number at the address 0000000000000110 is 0000000000000010. Thus, the ld instruction loads 0000000000000010 into the ac register.

You may find it difficult to read the long binary numbers in the preceding paragraph. So let's describe the action of the ld instruction again, but this time using hex notation: The ld instruction above—0006—loads the ac register from the memory location at the address 6. This location contains 0002. Thus, the ld instruction loads 0002 into the ac register. The memory location at the address 6 is unaffected.

The next instruction is an add instruction. This instruction adds a copy of the number in the memory location at the address 0007 hex to the ac register:

add opcode   12-bit memory address (007 hex)
0010   000000000111

The next instruction is a st (store) instruction:

st opcode   12-bit memory address
0001   000000000000

It contains the st opcode (0001). The remaining 12 bits in the instruction is the address of the memory location into which a copy of the number in the ac register is stored.

The last three instructions in our machine language program are *trap instructions*. When executed, they cause a transfer of control to the OS. The OS then performs some service, depending on the *trap vector* (the rightmost eight bits of the trap instruction). After performing the requested service, the OS returns control to the instruction following the trap instruction unless the trap instruction requests a halt. Here is the sequence of trap instructions we need in our program:

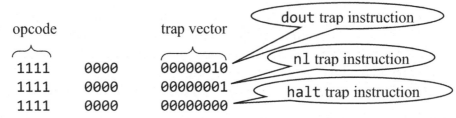

opcode		trap vector	
1111	0000	00000010	dout trap instruction
1111	0000	00000001	nl trap instruction
1111	0000	00000000	halt trap instruction

All the trap instructions have the same opcode (1111). However, the effect of each trap instruction differs and depends on its trap vector. For example, the first trap instruction above (whose vector is 00000010) displays in decimal the number in the ac register. The second trap instruction (whose vector is 00000001) moves the display cursor to the beginning of the next line. The third trap instruction (whose vector is 00000000) terminates the program. To distinguish the various trap instructions, we give each variation a unique name. For example, we give the trap instructions whose vectors are 00000010,

00000001, and 00000000 the names `dout` (decimal out), `nl` (new line), and `halt`, respectively.

To complete our program, we need two data numbers (2 and 3) following our six instructions and a third location to receive the sum of the addition:

```
0000000000000010
0000000000000011
0000000000000000
```

Here is the entire machine language program:

Address (hex)	Machine instruction (binary)	
0000	0000000000000110	(ld)
0001	0010000000000111	(add)
0002	0001000000001000	(st)
0003	1111000000000010	(dout)
0004	1111000000000001	(nl)
0005	1111000000000000	(halt)
0006	0000000000000010	(data)
0007	0000000000000011	(data)
0008	0000000000000000	(receives sum)

When executed, this program displays 5 (the sum of the two data words: 2 and 3). To try out this program, we have to create a file that contains the program in binary form. The easiest way to do this is to first create a text file that contains the program in hex form. We can represent each line of the program with a four-digit hex number. For example, the `dout` instruction,

```
1111 0000 0000 0010
```

in hex form is

```
f002
```

Let's create the file that contains the hex version of the program using any text editor (for example, `notepad` on Windows or `nano` on OS X). The file name extension should be ".`hex`". Suppose we create a text file named `e0201.hex` that contains the hex version of our program. We get

```
       e0201.hex
  0006    ; ld
  2007    ; st
  1000    ; add
  f002    ; dout
  f001    ; nl
  f000    ; halt
  0002    ; data
  0003    ; data
  0000    ; sum
```

We have added a comment to each line that describes the contents of that line. A comment starts with a semicolon and extends to the end of the line. Once we have a file containing the program in hex form, we can translate it to binary using the h2b program in the software package for this book. We invoke h2b (and the other programs in the software package for this book) from the *command line*. To get to the command line, start the `command prompt` program on Windows or the `Terminal` program on Mac OS X, Linux, and Raspbian.

To translate e0201.hex to binary, enter on the command line

      h2b e0201.hex   (on Windows)

or

      ./h2b e0201.hex   (on Mac OS X, Linux, or Raspbian)

h2b will then output the binary form of the program to the file named e0201.e. The output file name is the same as the input file name except for the extension—the output file name has the extension ".e" in place of ".hex". Now that we have our program in binary form, we can run it using the sim program (also in the software package for this book) which simulates the LCC. To run ex0201.e using the sim program, enter

      sim e0201.e   (on Windows)

or

      ./sim e0201.e   (on Mac OS X, Linux, and Raspbian)

In response, the sim program executes the program in e0201.e and displays the following:

```
Microlevel Simulator Version 1.0 Copyright (c) 2019 by Anthony J. Dos Reis
Opening machine code file e0201.e
Opening microcode file b.m
Opening log file e0201.log
========================================= output
5

==================================================
Machine code size:            9
Machine instructions executed: 6
Microcode size:               71
Microinstructions executed:   35
```

The output that the machine language program in e0201.e produces appears between the two rows of equal signs. Preceding the output on the display are the names of the files that the sim program uses. The sim program creates a ".log" file that is a record of what is displayed on the display screen. The ".log" file also includes a time stamp and your name. The first time you run the sim program, it will prompt you for your name. Thereafter, it will include your name in the ".log" files it creates. Following the output on the display are some statistics on the run. In this run, we can see that the size of our machine language program is 9, and 6 machine language instructions were executed. Each machine language instruction requires the LCC to execute a number of microinstructions. Thus, although only 6 machine language instructions were executed, 35 microinstructions were executed. The number of microinstructions executed is a measure of the execution time of the program.

Load Immediate Instruction

A `ld` instruction loads the `ac` register with a word from memory. Thus, it first fetches that word from memory. Then it loads it into the `ac` register. The `ldi` (load immediate) instruction also loads the `ac` register. But the word it loads comes from within the instruction itself. Thus, the `ldi` instruction does not have to perform the memory fetch operation that the `ld` instruction performs. For example, the following `ldi` instruction contains 000000001001 (009 hex) in its rightmost 12 bits:

$$\underbrace{1000}_{\text{ldi opcode}} \quad \underbrace{000000001001}_{\text{12-bit immediate operand}}$$

When the CPU executes this instruction, it extracts the 12 rightmost bits, zero-extends it to 16 bits to get 0000000000001001 (0009 hex), and then loads it into the `ac` register. The number loaded is in the instruction itself. Thus, as soon as the instruction is fetched by the CPU and placed in the `ir`, the operand is *immediately* available—it is sitting in the 12 rightmost bits of the `ir`. An operand that is in a machine language instruction is called an *immediate operand* because it is immediately available once the instruction is fetched and placed in the `ir`. The CPU treats the 12-bit immediate operand in a `ldi` instruction as an unsigned number. Thus, it cannot be a negative number. Because its only 12 bits in length, its range is only 0 to 4095 decimal (000 to fff hex).

Let's rewrite the program in `ex0201.hex` so that it uses the `ldi` instruction in place of the `ld` instruction. We get

```
      e0202.hex
    8002    ; ldi
    2006    ; add
    1007    ; st
    f002    ; dout
    f001    ; nl
    f000    ; halt
    0003    ; data
    0000    ; sum
```

The program is both shorter (because we do not need a data word for 2) and runs faster (because the `ldi` instruction is faster than the `ld` instruction) than the program in `e0201.hex`.

Strings

A string is a sequence of characters, each represented by a code. The code that the LCC uses is ASCII (American Standard Code for Information Interchange). ASCII represents each character with a seven-bit number which is usually zero-extended to 8 bits. Here are the ASCII codes for several characters:

```
'A':  01000001  (41 hex, 65 decimal)
'a':  01100001  (61 hex, 97 decimal)
'B':  00100010  (42 hex, 66 decimal)
'b':  01100010  (62 hex, 98 decimal)
'0':  00110000  (30 hex, 48 decimal)
'1':  00110001  (31 hex, 49 decimal)
' ':  00100000  (20 hex, 32 decimal)
'\n': 00001010  (0A hex, 10 decimal)
'\r': 00001101  (0D hex, 13 decimal)
```

The uppercase letters are assigned numbers in ascending order, starting from 41 hex. The lowercase letters are assigned numbers in ascending order, starting from 61 hex. The code for each uppercase letter differs from the code for the lowercase for that letter in only bit 5: In the uppercase letter, bit 5 is 0; in the lowercase letter, bit 5 is 1. The digits are assigned numbers in ascending order starting from 30 hex. '\n' is the *newline character*. '\r' is the *return character*. On systems other than Windows, just the '\n' character usually marks the end of each line in a text file. On Windows systems, the two-character sequence, '\r', '\n', marks the end of each line of a text file. For example, suppose a text file on a Windows system contains the following text:

AB
0 1

It is represented with the following sequence of ASCII codes (given in hex):

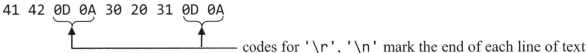

codes for '\r', '\n' mark the end of each line of text

The newline, return, and space characters are called *whitespace* since they do not produce a displayable character in print on a paper. You see just the white background—hence the name "whitespace."

On the LCC, a string of characters is represented by the sequence of their ASCII codes, each occupying *one word*, followed by the *null character* (the character represented by all zero bits). For example, if the string "AB" is in memory starting at location 200 hex, the locations of memory starting at address 0200 contain the following codes:

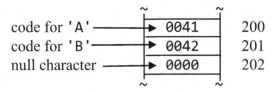

The null character marks the end of the string.

It is a good idea to memorize the ASCII codes for 'A', '0', space, '\n', and '\r'. That will allow you to quickly identify and decipher strings in their binary or hex representations.

To display a string on the display monitor, we use the `trap` instruction whose trap vector (i.e., the rightmost 8 bits in the trap instruction) is 00000110 (06 hex). When this `trap` instruction is executed, the `ac` register should have the address of the first character of the string. The `trap` instruction displays all the characters in memory starting from the address in the `ac` register until it reaches the null character. The name of this particular trap instruction is `sout` (for "string out").

Let's write a program that displays the string "hi\n". The ASCII codes for 'h', 'i' and '\n' are 01101000 (68 hex), 01101001 (69 hex), and 00001010 (0a hex), respectively. At the bottom of our program, following the `halt` instruction, we need three words containing these codes followed by the null character. When the `sout` instruction is executed, the `ac` register should contain the address of the string. We use a `ldi` instruction to load the `ac` register with that address:

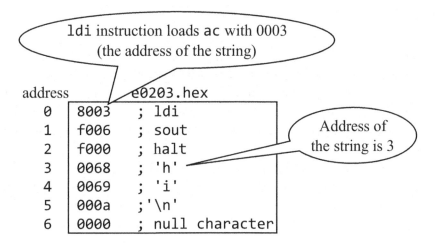

Branch Instructions

Machine instructions are normally executed serially in order of increasing memory address. For example, after the machine instruction at address 0 is executed, the machine instruction at address 1 is executed, then the machine instruction at address 2, and so on. However, a branch instruction can change this execution pattern. It does this by loading a new value into the `pc` register. Recall from Fig. 2.2, that the CPU executes a loop that repeatedly

1. fetches the instruction that the `pc` register points to
2. increments the `pc` register (so it is pointing to the next instruction in memory)
3. decodes the opcode (i.e., determine the opcode)
4. executes the instruction is just fetched.

Thus, when the `pc` register contains 2, the CPU fetches the instruction that is in memory at address 2 (assume a branch instruction is there), increments the `pc` to 3, and then executes the branch instruction. However, when the branch instruction is executed, it can change the address in the `pc` register. For example, it could change the address in the `pc` register to 0. Then when the CPU proceeds to fetch the next instruction, it fetches the instruction at address 0—not at address 3—because the `pc` register now contains 0. For example, consider the following program:

immediate operand zero-extended to 16 bits and then loaded into the `ac` register

```
1000 000000000101   ; ldi
1111 000000000010   ; dout
1100 000000000000   ; br (unconditional branch)
```

address zero-extended to 16 bits and then loaded into the `pc` register

In hex format, it is

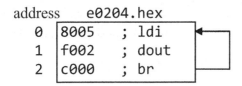

The first instruction (the ldi instruction) loads the ac register with 5. The second instruction (the dout instruction) displays the number in the ac register (5) in decimal. The third instruction (the br instruction) branches back to the address 0 (by loading the pc register with 0). Thus, the three instructions are again executed. Each time the br instruction is executed, the CPU branches back to address 0. Thus, the three instructions make up an *infinite* (i.e., never ending) *loop*. Try running this program. You will see it fills the screen with a never-ending stream of 5's.

The br instruction is an *unconditional branch*. That is, when it is executed the branch always occurs. The basic instruction set also has two conditional branch instructions: brz (branch on zero) and brn (branch on negative). The brz branches only if the value in the ac register is zero. The brn instruction branches only if the value in the ac register is negative. For example, in the following program, the ldi instruction initializes the ac register with 3. The first time the brz instruction is executed, it does not branch because the value in the ac register is not zero. The dout instruction then displays 3. The sub instruction subtracts 1 (the constant at the address 0006) from the ac register. So now the ac register contains 2. The br instruction then unconditionally branches back to the brz instruction. The same sequence of instructions, starting with brz instruction, is executed, but this time with 2 in the ac register (so 2 is displayed). The sequence is executed a third time with 1 in the ac register (so 1 is displayed). However, when the brz instruction is executed the fourth time, the ac register contains 0. Thus, the brz instruction branches to the halt instruction, terminating the program. The net effect of the program is to display "321" and then halt.

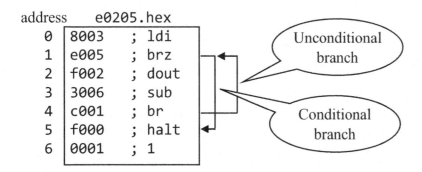

Trap Instructions

We have already seen four trap instructions: dout (decimal out), nl (newline), halt, and sout (string out) corresponding to the trap vectors 02 hex, 01 hex, 00 hex, and 06 hex, respectively. For example, here is the halt trap instruction in binary:

```
opcode (f hex)  trap vector (00 hex)
   1111         0000 00000000
```

This instruction written in hex is `f000`. The trap instructions for `dout`, `nl`, and `sout` written in hex are `f002`, `f001`, and `f006`, respectively. Here is a complete list of the trap instructions:

Trap instruction	Name	Function
f000	halt	Terminate execution
f001	nl	Output newline to display
f002	dout	Display signed number in ac in decimal
f003	udout	Display unsigned number in ac in decimal
f004	hout	Display number in ac in hex
f005	aout	Display character in ac
f006	sout	Display string ac points to
f007	din	Read decimal number from keyboard into ac
f008	hin	Read hex number from keyboard into ac
f009	ain	Read character from keyboard int ac
f00A	sin	Read in string to address in ac register
f00E	bp	Breakpoint

Let's write a program that reads in a decimal number and displays its value in hex. Here's what you should see on the screen when you run the program if you input the decimal number 30:

```
30
1E
```

Of course, a properly designed program should prompt the user before reading in a number so the user is notified that the program is waiting for input. It should also label the output. A session should look like this:

```
Enter decimal number
30
Number in hex = 1E
```

But let's keep our program simple for now by omitting the label on the output and the prompt.

To read in a decimal number into the `ac` register, we use the `din` trap instruction. The `din` (decimal in) instruction reads the decimal number entered on the keyboard, converts it to a 16-bit binary number, and loads it into the `ac` register. Then to display the number in the `ac` register in hex, we use the `hout` trap instruction. It converts the number in the `ac` register to hex and displays it on the screen. Here is the program:

e0206.hex
```
f007    ; din
f004    ; hout
f000    ; halt
```

If you run this program, be sure to enter a decimal number after you start it. It does not prompt for the decimal number so the `sim` program will just wait for it indefinitely without giving you any notification that it is waiting for input.

Problems

1) Write and run a machine language program in binary that adds 1, 200, and 500 and displays the sum in decimal and hex.

2) Write and run a machine language program that adds −1, −200, and −500 and displays the sum in decimal and hex.

3) Rewrite and run the program given `ex0206.hex` so that it prompts the user and labels the output.

4) Write and run a machine language program that reads in a character (use `ain`) and displays its ASCII code in decimal and hex. Test your program with `'A'`, `'a'`, and `'0'` (when entering a character in response to the `ain` instruction, do not enclose the character in quotes). Also run your program and do not enter any character. Just immediately hit the Enter key (which inputs the newline character).

5) Write and run a machine language program that reads in a decimal number (use `din`). Output what is read in with `dout`, `udout`, and `hout`. Test your program with −1. Display each value on a separate line.

6) Write and run a machine language program that displays the numbers 1 to 10. Use a loop.

7) Write and run a machine language program without using `sout` that displays your name.

8) Write and run a machine language program that reads in a positive decimal number and then displays "hello" that number of times.

9) Write and run a machine language program that reads in two decimal numbers and displays "bigger" if the first number is bigger than the second, "equal" if the two numbers are equal, or "smaller" if the first number is smaller than the second. Test you program with 5 and 3, 3 and 5, 3 and 3. Also test your program with 32767 and −1. *Hint*: Subtract and test the result with the `brn` and `brz` instructions.

10) Write and run a machine language program the computes and displays the sum of the first 15 positive odd numbers.

3 Assembly Language

Introduction

Assembly language is a symbolic form of machine language. An assembly language instruction consists of symbols that represent the binary fields of a machine language instruction. Because it is symbolic, assembly language is easier to read, write, modify, and debug.

Here is a machine language program from chapter 2 (it loads the `ac` register from address 6, adds from address 7, stores the sum in at address 8, and displays the sum in the `ac` register) along with its corresponding assembly language program (which is in the file `e0301.a`). Note that for files containing an assembly language program, we use the file name extension ".a":

```
Machine Language              Corresponding
    Program               Assembly Language program
                                 e0301.a
0000000000000110         ld x     ; load ac with value at x
0010000000000111         add y    ; add value at y to ac
0001000000001000         st z     ; store value in ac at z
1111000000000010         dout     ; display value in ac in decimal
1111000000000001         nl       ; move cursor to next line
1111000000000000         halt     ; stop execution
0000000000000010      x  .fill 2  ; data
0000000000000011      y  .fill 3  ; data
0000000000000000      z  .fill 0  ; data
```

If a semicolon appears in a line of an ".a" file, everything from the semicolon to the end of that line is a comment. A *comment* is information simply for our edification—it does not contribute to the specification of the program.

Consider the first instruction—the `ld` instruction—in the two programs above. The assembly language program uses "`ld`" in place of the binary opcode 0000. We call the names like `ld` that we use in place of opcodes *mnemonics* because they are easy to remember ("mnemonic" means "aiding memory"). In place of the address in the `ld` machine language instruction, we use the label x. Think of x as a symbolic address. It represents the address of the line in the program that starts with the label x:

```
x       .fill 2
```

The directive `.fill 2` indicates that 2 in binary should "fill" (i.e., completely occupy) the memory location corresponding to this line. That is, the corresponding word in the machine language program should contain 0000000000000010. We do not call `.fill` a mnemonic because it is not the name of an opcode. Instead, we call it a *directive* because it directs us to do something when translating the program—namely, to replace the line with the specified data converted to binary. Similarly, in place of the address in the `add` instruction, we use the label y, which represents the address of the line that starts with the label y. That line has a `.fill` directive that "fills" the memory location corresponding to that line with 3 in binary.

Rule: Mnemonics and directives but not labels are *case insensitive*. That is, they can be in either upper or lower case in an assembly language program. Thus, add and ADD are equivalent but the labels x and X are not.

Rule: Labels must start in column 1. Mnemonics and directives can start in any column except column 1. Comments can start in any column.

In a .fill directive, we can specify a number in decimal or hex, a character constant (a character within single quotes), or a label. A character constant is translated to its ASCII code. A label is translated to its corresponding 16-bit address. For example, if the following statements appear in a program, and the address of the line that starts with the label b is 0124 hex, then the memory location corresponding to the line that starts with the d label would contain 0000000100100100 (0124 hex).

```
a          .fill 3      ; translated to 0000000000000011 (3 in binary)
b          .fill 0xf1   ; translated to 0000000011110001 (f1 hex in binary)
c          .fill 'A'    ; translated to 0000000001000001 (ASCII code for 'A')
d          .fill b      ; translated to the addr of line that starts with b
```

One of the great benefits of using labels in place of addresses is that insertions and deletions do not require any changes in labels. For example, suppose we want the modify the programs in e0301.a so that it displays the computed sum in both decimal and hex. In the assembly language program, we simply insert

```
      hout         ; displays value in ac in hex
      nl           ; move cursor to next line
```

just before the halt instruction. No other changes are required. But in the machine language program, the insertion of the machine language instructions hout and nl *changes the addresses of the three data words* at the bottom of the program. Thus, the addresses in the ld, add, and st instructions have to be adjusted so that they correspond to the new addresses of the data. In a large program with many instructions that address the data, a *single insertion or deletion might require hundreds of additional modifications* to correct for the altered addresses of the data. Clearly, this would be a clerical nightmare.

Assembly language instructions that have label operands can be written with the actual addresses specified in decimal or hex in place of the label operands. For example, the ld instruction in e0301.a can be written this way:

```
      ld 6    ; load the value at address 6
```

instead of this way:

```
      ld x    ; load the value at the label x
```

These instructions are translated by the assembler to the same machine code. However, if we use addresses instead of labels, we would incur the problem of having to adjust the addresses if we make insertions or deletions in the program. For this reason, we will always use label operands rather than addresses.

Rule: A label must start with "_", "$", or "@", or a letter. After the first character, the digits 0 to 9 are allowed in addition to "_", "$", "@", and letters.

Assembly language, of course, is not machine language. It cannot be executed directly by the CPU. Before an assembly language program can be executed, it must be translated to machine language. It is tedious but not difficult to translate assembly language programs to machine language by hand. But an easier and more reliable approach is to use a program—called an *assembler*—to do the translation for us. The `basic` program (in the software package for this book) is an assembler for the basic instruction set. To assemble the program in `e0301.a`, enter on the command line

      `basic e0301.a`   (on Windows)

or

      `./basic e0301.a`   (on OS X, Linux, or Raspbian)

The assembly process consists of two passes over the program in the input file. During the second pass, the machine code is outputted to the file `e0301.e`. This file has the extension ".e" (which indicates the file holds an executable machine language program) and the same base name as the input file.

The `basic` program also produces a ".lst" file with the same base name as the input file. Thus, for the input file `e0301.a`, it produces the file `e0301.lst`. This file contains the *source program* (i.e., the "source" of the assembly process—the assembly language program) along with the corresponding machine code and locations in hex format. Here is the `e0301.lst` file produced by the `basic` program:

```
Basic Assembler Version 1.0         Thu May  2 17:50:05 2019
DosReis Anthony J

Header

A 0000
A 0001
A 0002
C

Loc    Code         Source Code

0000   0006         ld x     ; load ac with value at x
0001   2007         add y    ; add value at y to ac
0002   1008         st z     ; store sum in ac in z
0003   f002         dout     ; display value in ac in decimal
0004   f001         nl       ; move cursor to next line
0005   f000         halt     ; stop execution
0006   0002 x       .fill 2  ; data
0007   0003 y       .fill 3  ; data
0008   0000 z       .fill 0  ;
```

To run the executable program in `e0301.e` produced by the `basic` assembler, we use the `sim` program (in the software package for this book). To run `sim`, on the command line specify `sim` and the name of the executable file produced by the `basic` assembler:

`sim e0301.e`   (on Windows)
   or
`./sim e0301.e`   (on OS X, Linux, or Raspbian)

You will then see on the display screen

```
Microlevel Simulator Version 1.0 Copyright (c) 2019 by Anthony J. Dos Reis
Opening machine code file e0301.e
Opening microcode file b.m
Opening log file e0301.log
======================================== output
5


==================================================
Machine code size:            9
Machine instructions executed: 6
Microcode size:              71
Microinstructions executed:  35
```

The first few lines displayed by `sim` indicate the files it is using or creating. Notice that for this example, `sim` is using the microcode file `b.m`. This is the microcode file that defines the basic instruction set. Any executable file created by the `basic` assembler will automatically trigger the use the microcode in `b.m`. `sim` also produces a ".log" file that is a time-stamped record of what you see on the display when you run `sim`.

ldi Instruction

In an assembly language `ldi` instruction, the operand can be a decimal or hex constant, a character constant, or a label. For example, consider the following program:

```
  ; e0302.a
0            ldi 15    ; immediate operand is 00f hex
1            dout      ; displays 15
2            ldi 'A'   ; immediate operand is 041 hex (ASCII code for 'A')
3            hout      ; displays 41
4            dout      ; displays 65
5            aout      ; displays A
6            ldi x     ; immediate operand is 009 hex (address of x)
7            dout      ; displays 9
8            halt
9 x          .fill 5
```

In the first `ldi` instruction, the operand is a decimal constant. The assembler puts this constant in binary form into the immediate field (the 12 rightmost bits) of the `ldi` machine instruction. Because the immediate field is 12 bits wide, it can accommodate constants in the range of 0 to 4095 decimal (0 to fff hex).

In the second `ldi` instruction, the operand is a character constant. It is translated to its ASCII code. Thus, the immediate field of this instruction contains 000001000001 (041 hex), the ASCII code for the letter A. When executed, this `ldi` instruction loads the `ac` register with value in its immediate field, zero-extended to 16 bits. Thus, it loads the `ac` register with 0000000001000001 (0041 hex), the ASCII code for A. The hout (hex out) and dout (decimal out) instructions at addresses 3 and 4 then display, respectively, 41 and 65 (65 is the decimal equivalent of 41 hex). The aout instruction at address 5 displays the character whose ASCII code is in the ac register. Thus, it displays the letter A.

In the third ldi instruction, the operand is a label. The assembler always translates labels to their corresponding addresses. Thus, x is translated to 000000001001 (009 hex), which is the address of x. Thus, this ldi instruction loads the address of x into the ac register.

.Blkw and .Stringz Directives

In addition to the .fill directive, two directives that we will use frequently are .blkw and .stringz. The .blkw directive reserves and initializes to zero a block of words in memory. For example, the following .blkw directive reserves and initializes 100 words in memory:

```
buffer      .blkw 100       ; 100 is a decimal constant
   or
buffer      .blkw 0x64      ; 0x64 is a hex constant
```

This single directive is equivalent to

```
buffer      .fill 0  ⎫
            .fill 0  ⎬  100 .fill directives
               ⋮     ⎪
            .fill 0  ⎭
```

If a program inputs a string from the keyboard, it must have a memory area big enough to receive the string. We call a memory area in a program that receives data on an input operation or provides data on an output operation a *buffer*. If the maximum length of an inputted string is *n*, then the buffer should be at least *n*+1 words long (the "+1" is for the null character at the end of the string). Thus, the buffer above can accommodate strings with lengths up to 99 characters.

We use the .stringz directive to create a null-terminated string. For example, the .stringz directive

```
greeting    .stringz "Hello, world"
```

creates the sequence of ASCII codes corresponding to the characters in the string followed by the null character. Recall that the null character is a word that contains zero. The "z" (for zero) at the end of the name of this directive is an indication that this directive creates a string with a zero word (i.e., the null character) at the end. Recall that a character constant is delimited with *single* quotes (for example, 'A'). Note that string constants are delimited with *double* quotes.

Let's now write an assembly language program that reads in a string from the keyboard and echoes it to the display. We will use the sout trap instruction (to display a prompt message and to echo the inputted string) and the sin trap instruction (to input the string). For both instructions, the ac register must have the appropriate address: For the sout instruction, the ac register must have the address of the string to be displayed; for the sin instruction, the ac register must have the address of the buffer that is to receive the string. We will use the ldi instruction to load the ac register with the required addresses.

```
; e0303.a
0               ldi prompt          ; load ac with address of prompt message
1               sout                ; display prompt message
2               ldi buffer          ; load ac with address of buffer
3               sin                 ; read string from keyboard into buffer
4               sout                ; echo string to display
5               halt
6   prompt      .stringz "Enter string\n"
14  buffer      .blkw 100
```

The addresses on the left are not part of the program. The addresses are in hex so that last line is at address 14 hex (20 decimal). We include the address of each line in our program listings so we can easily refer to specific lines in the program. The assembler translates "prompt" in the `ldi` instruction to its corresponding address (006). Thus, 006 is the immediate value in the `ldi` instruction. When the `ldi` instruction is then executed, its immediate value is zero-extended to 0006 and loaded into the `ac` register. Thus, the `ac` register receives 0006, which is the address of the prompt message created by the `.stringz` directive.

The `sout` instruction displays the message whose address is in the `ac` register (0006), which prompts the user to enter a string. The `ldi` instruction at address 2 loads the `ac` register with the address of the 100-word buffer created with the `.blkw` directive. The `sin` instruction then reads in a string entered on the keyboard into memory starting at that address. After the `sin` instruction is executed, the address of the buffer is still in the `ac` register. Thus, when the `sout` instruction at address 4 is executed, it echoes the string just inputted to the display. For example, if "hello" is entered on the keyboard when this program is run, the program will echo back the string entered. Thus, you will see on the screen two occurrences of "hello":

```
hello
hello
```

Branch Instructions and Loops

Instructions are normally executed in memory order. For example, after the instruction at address 0000 is executed, the instruction at address 0001 is executed, then the instruction at 0002, and so on. However, a branch instruction can alter this pattern of instruction execution. When executed, a branch instruction can load the `pc` register with an address, causing the CPU to "branch" to that address. For example, if the branch instruction at address 0000 loads the `pc` register with 0030, then the next instruction the CPU executes will be the instruction at the address 0030—not the instruction at 0001.

In assembly language, a branch instruction consists of the mnemonic that specifies the opcode and the label on the line of code to branch to. For example, the following branch instruction branches to the line of code that starts with the label "sub":

```
    br sub
```

Recall that in addition to the unconditional branch instruction `br`, there are the conditional branch instructions: branch on negative (`brn`) and branch on zero (`brz`) instructions. The conditional branch instructions branch depending on the value in the `ac` register. For example, the `brz` instruction branches if the value in the `ac` register is zero. If the value in the `ac` register is not zero, then execution continues with the instruction in memory that follows the `brz` instruction.

Here is the looping program in `e0205.hex` in hex form from chapter 2 along with its assembly language equivalent in `e0304.a`:

```
address     e0205.hex                           e0304.a
   0    8003   ; ldi                    ldi 3      ; init ac with 3
   1    e005   ; brz          loop      brz done   ; branch when ac = 0
   2    f002   ; dout                   dout       ; display value in ac
   3    3006   ; sub                    sub @1     ; subtract 1 from ac
   4    c001   ; br                     br loop    ; branch back to loop
   5    f000   ; halt         done      halt       ; stop execution
   6    0001   ; 1            @1        .fill 1    ; constant 1
```

The `ldi` instruction initializes the `ac` register with 3. Each pass through the loop displays the value in the `ac` register and then decrements it by 1. At the beginning of the fourth pass through the loop, the `ac` register contains 0. Thus, on the fourth pass, the `brz` instruction at address 1 branches to the `halt` instruction at the label `done`. This program displays "321".

The label `@1` in `e0304.a` is a legal label (recall that labels can start with @, $, _, or a letter). Alternatively, we could have used the label `x` in place of `@1`. But `@1` is better because it indicates the constant at that label. When we see the label `@1` in an instruction, we know immediately that it labels a `.fill` directive with the constant 1. We do not have to search for its `.fill` directive (a time-consuming process in a big program) to determine its corresponding constant. Labels on `.fill` directives with non-negative constants should be prefixed with "@". Labels on `.fill` directives with negative constants should be prefixed with "@_". We cannot use a minus sign within a label. In its place, we use the underscore character. For example, the `.fill` directive for −5 should be

```
@_5         .fill -5
```

Accessing the Stack

A *stack* is a linear data structure that is accessed from one side only. The side accessed is called the *top* of the stack. The operation that adds an item to the top of the stack is called a *push*; the operation that removes the item on top of the stack is called a *pop*.

In an assembly language program, the stack is located at the top of memory (i.e., starting at the address 65535). As items are pushed onto the stack, the stack grows in the downward direction (i.e., toward locations with smaller addresses). Thus, on a push, the address of the top of the stack decreases by one; on a pop the address increases by one. To keep track of the top of the stack, we use `sp` register. It always "points to" (i.e., contains the address of) the top of the stack.

When several items have been pushed onto the stack, we frequently want to access them without popping them. We do this with the relative instructions (`ldr`, `str`, `addr`, and `subr`). For example, suppose the stack and the `sp` register are configured as follows:

30   Chapter 3: Assembly Language

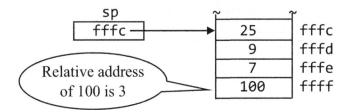

25 is in the location sp points to. Thus, its relative address (i.e., its address relative to the location sp points to) is 0. The value 9 has the relative 1 because it is one word higher in memory than the location sp points to. Similarly, 7 has the relative address 2, and 100 has the relative address 3. To load the ac register with the word at relative address 3, we use a ldr instruction in which specify the relative address 3:

    ldr 3

To then add the value at relative address 2 (which is 7), we use

    addr 2

To then subtract the value at relative address 1 (which is 9), we use

    subr 1

To then store the value in the ac register into the location corresponding to relative address 0, we use

    str 0

This instruction stores $100 + 7 - 9 = 98$ in the location that sp points to, overlaying the 25 that is there.
    Sometimes we want to reserve some slots on the stack into which we later store values. To do that, we simply decrement the value in sp using the asp instruction. For example, to reserve two slots on the stack, we use

    asp -2

This instruction adds the value it specifies (−2 in this example) to the address in the sp register. Thus, its effect is to decrement the address in the sp register by 2, thereby reserving two slots on the stack. The stack then looks like this:

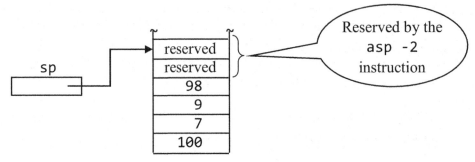

We now have two reserved location on the stack that we can use later. We can now store values in them. For example, to store 11 in the lower reserved location (the one that `sp` points to), we use

```
ldi 11      ; load the ac register with 11
str 0       ; store it at relative address 0
```

Note that by reserving two slots on the stack, the relative addresses on each item on the stack increase by two. For example, the relative address of 100 is now 5. Before reserving the two slots, it was 3.

The CPU treats the 12-bit field in the `asp` instruction as a *signed* number. Thus, it can hold numbers from −2048 to +2047 decimal. We use a negative number in an `asp` instruction to reserve words on the stack; we use positive numbers to remove items on the stack. For example, to reserve 10 words, use

```
asp -10
```

To remove the top 10 items from a stack, use

```
asp 10
```

Calling Subroutines

A *subroutine* is a segment of code in a program that performs a specific task. It is executed when an instruction in the program *calls* it. When finished performing its task, a subroutine executes a *returning instruction* that loads the `pc` register with the address of the instruction in memory that follows the calling instruction, causing the CPU to return to the calling module and execute instructions from there:

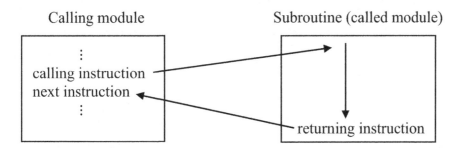

We refer to the module that contains the calling instruction the *caller* or the *calling module* and the subroutine as the *called module*.

The calling instruction causes the CPU to branch to the subroutine (by loading the `pc` register with the address of the beginning of the subroutine). But it also saves the address of the instruction that follows the calling instruction. This address—the *return address*—is needed by the returning instruction. When executed, the returning instruction loads the return address into the `pc` register, causing a return to the instruction that follows the calling instruction in the calling module.

The calling instruction in the basic instruction set is the `call` instruction. Here is the call instruction that calls a subroutine that starts with the label `sub`:

```
call sub
```

Recall from the preceding chapter that before the CPU executes an instruction, it increments the `pc` register. Thus, when a `call` instruction is executed, the `pc` register has the address of the next instruction. This address *is the return address*. The `call` instruction saves this address by pushing it onto the stack. After the `call` instruction saves the return address on the stack, it loads the `pc` register with the address of the subroutine (which is in the rightmost 12 bits of the `call` instruction), which causes the CPU to branch to the subroutine. When the subroutine subsequently executes the `ret` (return) instruction, the `ret` instruction pops the return address into the `pc` register, causing a branch back to the caller—specifically to the instruction that follows the `call` instruction.

Let's look at a program that consists of a main module and a subroutine. The main module calls the subroutine twice. The subroutine displays "hello" and returns to the main module. Here is the program:

```
  ; e0305.a
0 main         call sub      ; saves return address (0001) and branches to sub
1              call sub      ; saves return address (0002) and branches to sub
2              halt
3 ;================
4 sub          ldi msg
5              sout
6              ret           ; pop return address into pc
7 msg          .stringz "hello\n"
```

Note that the return address for the first call of `sub` is the address of the second call. For the second call of `sub`, the return address is the address of the `halt` instruction. When executed, this program displays "hello" twice, once for each call of `sub`. Note that the label on the subroutine, `sub`, is also the mnemonic for the subtract instruction. However, this use of `sub` does not cause a problem. The assembler determines from the way `sub` is used that it is a label rather than the subtract instruction mnemonic.

The *entry point* of a program is the location in a program where execution should start. By default, the entry point of a program is the physical beginning of the program. If, however, the entry point is not the physical beginning of a program, we must specify it with a `.start` directive. For example, suppose we reverse the order of the `main` and `sub` functions in the preceding program. We get

```
  ; e0306.a
0 sub          ldi msg
1              sout
2              ret
3 msg          .stringz "Hello\n"
a ;================
a main         call sub      ; saves return address (0001) and branches to sub
b              call sub      ; saves return address (0002) and branches to sub
c              halt
d              .start main   ; specifies entry point
```

(Entry point is here — at `main`)

Because the entry point of this program is `main`, and `main` is not at the physical beginning of the program, we have to indicate the entry point with a `.start` directive. The `.start` directive at the end of the program,

```
        .start main
```

indicates that the entry point of the program is at the label `main`.

A `.start` directive can appear anywhere in the program. For example, it can appear between the two `call` instructions in the program above. A `.start` directive is not translated to machine code. Thus, if we place the `.start` directive in between the two `call` instructions in the program above, the two `call` instructions would still be right next to each other in memory.

Passing Arguments Via the Stack

Let's now write a program that consists of a `main` function and a subroutine. The `main` function passes two arguments, 9 and 8, to the subroutine via the stack. The subroutine accesses them on the stack, subtracts 8 from 9, leaving the difference in the `ac` register. On return to the main function, `main` removes the two arguments from the stack and then displays the difference returned in the `ac` register. Here is the program:

```
  ; e0307.a
0 main        asp -1    ; three inst sequence that pushes 9 onto the stack
1             ldi 9
2             str 0
3             asp -1    ; three inst sequence that pushes 8 onto the stack
4             ldi 8
5             str 0
6             call sub  ; call sub (push return address and branch to sub)
7             asp 2     ; remove parameters from stack
8             dout      ; display value returned by sub
9             nl        ; move cursor to the next line
a             halt
  ;====================
b sub         ldr 2     ; load 9
c             subr 1    ; subract 8 from 9
d             ret       ; return with 1 in ac register
```

The three-instruction sequence that starts at address 0 pushes 9 onto the stack. The `asp` instruction reserves one word on the stack, the `ldi` instruction loads the `ac` register with 9, and the `str` instruction stores the value in the `ac` register into the reserved slot on the stack. At addresses 3, 4, and 5, we repeat to sequence but with the constant 8. The `call` instruction at address 6 pushes the return address (0007) in the `pc` register onto the stack and then branches to the subroutine. On entry into the subroutine, the stack looks like this:

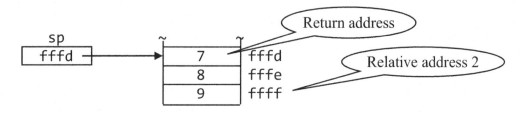

The relative address of 9 is 2; the relative address of 8 is 1. Thus, the instructions

```
        ldr 2
        subr 1
```

load the `ac` register with 9 and then subtracts 8 from it. Thus, when the `ret` instruction is executed, the difference 1 is in the `ac` register.

The `ret` instructions pops the return address off the stack into the `pc` register, causing a return to the `asp` instruction at the address 7. The `asp` instruction pops the two parameters previously pushed onto the stack by adding 2 to the `sp` register. `main` then displays the value in the `ac` register (1), moves the cursor to the beginning of the next line, and then terminates.

Note that the calling function (`main` in this program) creates the parameters passed to the subroutine by pushing the values of the arguments (9 and 8) onto the stack. It is the responsibility of the calling function to ultimately remove these parameters. It does so immediately after the subroutine returns to the calling function.

Basic Instruction Set Summary

Let's summarize the instructions in the basic instruction set. A summary is also in Appendix B and in the software package for this book in the file `basic.txt`. You may find it helpful to print out a copy of this file and have it handy as you write programs using the basic instruction set.

In the summary of the basic instruction set that follows, we let x represent the unsigned number in the 12 rightmost bits in the instructions like `ld` and `st`, and we view memory as an array named `mem`. We can then describe the action of instructions using C-like statements. For example, the effect of the `ld` instruction is described with the following C-like statement:

```
        ac = mem[x];
```

That is, a `ld` instruction "assigns" the word at the address x in memory to the `ac` register.

In our descriptions, x represents an unsigned number. But s represents a signed number—the signed number in the 12 rightmost bits of the `asp` instruction. Thus, s can be any number from −2048 to +2047. y represents the unsigned number in the eight rightmost bits of the `trap` instruction.

The `call` instruction pushes the address in the `pc` register onto the stack before it loads the `pc` register with a new address. This push operation can be succinctly captured using our C-like notation. In a push operation, the `sp` register is first decremented. It is then used to provide the address of the memory location into which the `pc` is stored. Here is the C-like notation that captures this sequence of operations:

```
        mem[--sp] = pc;        // push pc
```

In this statement, the decrement operator "--" precedes "sp." It indicates that the `sp` register is decremented *before* the address it contains is used in the memory store operation.

The `ret` instruction can similarly be described using our C-like notation:

```
        pc = mem[sp++];        // pop into pc
```

The increment operator, "++", follow the `sp` register. It indicates that the `sp` register is incremented *after* it the address it contains is used to access memory. It performs a pop operation.

Opcode	Format		Description
0	ld	x	ac = mem[x];
1	st	x	mem[x] = ac;
2	add	x	ac = ac + mem[x];
3	sub	x	ac = ac - mem[x];
4	ldr	x	ac = mem[sp + x];
5	str	x	mem[sp + x] = ac;
6	addr	x	ac = ac + mem[sp + x];
7	subr	x	ac = ac - mem[sp + x];
8	ldi	x	ac = x;
9	asp	s	sp = sp + s;
A	call	x	mem[--sp] = pc; pc = x; *(Push operation)*
B	ret		pc = mem[sp++]; *(Pop operation)*
C	br	x	pc = x;
D	brz	x	if (ac == 0) pc = x;
E	brn	x	if (ac < 0) pc = x;
F	trap	y	see below

halt	or trap 0	Terminate program
nl	or trap 1	Output nl character
dout	or trap 2	Output number in ac as signed decimal
udout	or trap 3	Output number in ac as unsigned decimal
hout	or trap 4	Output number in ac in hex
aout	or trap 5	Output character in ac
sout	or trap 6	Output string pointed to by ac
din	or trap 7	Input decimal number into ac
hin	or trap 8	Output hex number into ac
ain	or trap 9	Input character into ac
sin	or trap 10	Input string to address in ac
bp	or trap 14	Breakpoint

x: bits 0 to 11 in machine instruction zero-extended to 16 bits
s: bits 0 to 11 in machine instruction sign-extended to 16 bits
y: bits 0 to 7 in machine instruction zero-extended to 16 bits
ac: accumulator register
pc: program counter register
sp: stack pointer register

Directives: .blkw, .fill, .start, and .stringz

Problems

1) Write an assembly language program that displays your name.

2) Write an assembly language program that displays your name 20 times. Use a loop.

3) Write a program that repeatedly prompts for and reads in decimal numbers until a negative number is entered, at which point your program should display the sum of all the numbers previously entered. Test your program by entering 1, 2, 3, −1 (the sum displayed should be 6).

4) Write a program that prompts for and reads in a positive decimal number and then displays the sum of all the integers from 1 to the number entered. Test your program by entering 10.

5) Write an assembly language program that reserves a slot on the stack (by decrementing the `sp` register), pushes 1 and 2 onto the stack, and then calls a subroutine. Your subroutine should add the two numbers it is passed, store the sum in the reserved slot on the stack, and then return to the caller. The caller should then remove the two parameters from the stack, display the sum that is now on top of the stack, and finally remove the sum from the stack and terminate.

6) Write an assembly language program that displays a table of ASCII codes from 32 to 127 and their corresponding characters.

7) Write a program that works like the program in `e0206.hex` but prompts the user and labels the output.

8) Write an assembly language program in which the main module reserves one word on the stack (by decrementing `sp`) and calls a subroutine. Your subroutine should prompt for and read in a decimal number and store in the reserved word on the stack and then return to the caller. The caller should then display the number in the reserved word on the stack and then remove it from the stack. Test your program by entering −5.

9) Write an assembly language program that reads in 10 decimal numbers and displays the largest. Test your program by entering 3, 100, −30, −50, 101, 99, 0, −1, 5, 77. *Hint*: To determine if x is less than y, subtract y from x and examine the result. If the result is negative, x is less than y.

10) Write an assembly language program that prompts for and reads in a string. It should then display all and only the decimal digits in the string. Test your program with "A1b2C34".

11) Write an assembly language program that reads in a string and displays it with its characters in reverse order. Test your program by entering "hello". Use the stack.

12) Write an assembly language program that reads in a hex number and displays it in binary. Test your program by entering AB5D.

13) When the following program is executed, does it loop infinitely? If so, explain why. If not, explain why not. Assume the program is loaded into location 0.

```
            ld x
s           asp -1
            str 0
            br s
x           .fill 0xf000
```

14) Write an assembly language program that reads in a four-bit binary number (with spaces separating the bits) and displays its equivalent hex digit. Test your program by entering 1 1 0 1 for which your program should display "d".

15) Hand assemble the program in `e0302.a`. Give your answer in hex.

16) Hand assemble the program in `e0303.a`. Give your answer in hex.

17) Hand assemble the program in `e0304.a`. Give your answer in hex.

18) Hand assemble the program in `e0305.a`. Give your answer in hex.

19) Hand assemble the program in `e0306.a`. Give your answer in hex.

20) Hand assemble the program in `e0307.a`. Give your answer in hex.

4 Simple Digital Circuits

Introduction

In this chapter, we will study the simple digital circuits that are the basic building blocks of the LCC. We will study the two categories of digital logic: combinational (a circuit whose output depends only on its present inputs) and sequential (a circuit that has memory, and therefore its output can depend on its past as well as its present inputs).

Transistors

A *transistor* is an electrical switch. It can be *off*, in which case is does not conduct electricity, or *on*, in which case is does conduct electricity:

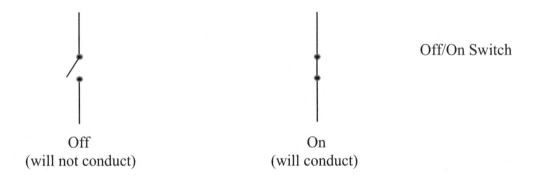

Off/On Switch

Off
(will not conduct)

On
(will conduct)

A transistor is similar to a switch on a wall that operates an overhead light. There is, however, one major difference. A light switch is mechanical. Because it has moving parts, its switching speed is very slow. A transistor, on the other hand, is electrical. Because it has no moving parts, its switching speed is very fast.

Transistors are made from silicon. Silicon is an element that is in abundant supply on the Earth (silicon is the principal component of sand). We call silicon a *semiconductor* because its electrical properties place it between conductors and insulators.

Among the several types of transistors, metal-oxide semiconductor (MOS) transistors are the most commonly used type in modern computers. MOS transistors come in two types: PMOS (positive metal-oxide semiconductor) and NMOS (negative metal oxide semiconductor). Most computer circuits with MOS transistors use complementary metal-oxide semiconductor (CMOS) technology. In this technology, both NMOS and PMOS transistors are used. By combining both types in a circuit, the power required by the circuit is minimized. MOS transistors have several properties that make them appropriate for computer circuits: low power consumption, high noise immunity, and high fanout (*fanout* is the number of circuits that the output of a single gate can drive). They, however, are generally slower that other types of transistors.

MOS transistors have three leads—the *source*, *gate*, and *drain*. An off/on switch exists between the source and drain. The status of this switch—on or off—is determined by the voltage applied to the gate. For example, here are schematic pictures of two NMOS transistors, one with 0 volts on its gate and one with 3 volts on its gate, along with their equivalent circuits:

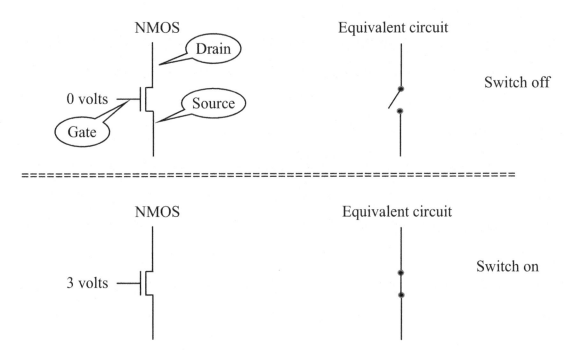

When 0 volts is applied to the gate of an NMOS transistor, the switch is turned off. When 3 volts is applied, the switch is turned on. A PMOS transistor works in the opposite way: When 0 volts is applied to the gate of a PMOS transistor, the switch is turned on. When 3 volts is applied, the switch is turned off:

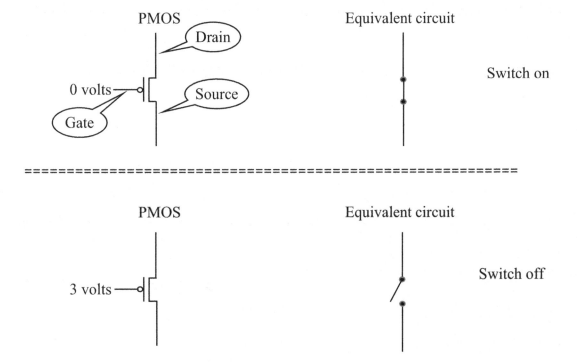

Note that the schematic representation for a PMOS transistor has a small "bubble" on the gate input. This bubble distinguishes it from a NMOS transistor, and it indicates that the transistor is activated (i.e., turned on) when a zero voltage is applied.

Two voltage levels are used on MOS transistors in a computer: typically, 0 volts and a low positive voltage (such as 3 volts). The two levels are used to represent the two values of a bit. For example, 0 volts can represent the bit 0, and 3 volts can represent the bit 1.

NOT Gate

A *NOT gate* (also called an inverter) is a digital circuit that inverts the input voltage. That is, if the input voltage is 3, the output voltage is 0; if the input voltage is 0, the output voltage is 3. Here is a NOT gate constructed with both a PMOS and a NPOS transistor (the equivalent circuit is for the NOT gate when 3 volts are applied to its input):

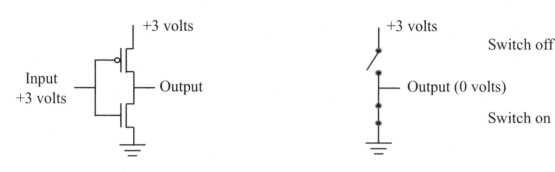

The symbol at the bottom of the circuit consisting of three, successively smaller horizontal lines represents ground (i.e., 0 volts). The top of the circuit is connected to +3 volts. If we apply 3 volts to the input of the circuit, the top transistor turns off (because it is PMOS) and the bottom transistor turns on (because it is NMOS). The effect is to connect the output directly to ground, setting it to 0 volts.

If, on the other hand, we apply 0 volts to the input, then the top transistor turns on and the bottom one turns off, connecting the output to +3 volts:

Here is a table that represents the input/output relationship of a NOT gate:

input	output
0 volt	3 volts
3 volts	0 volts

NOT Gate

If we represent a 1 bit with 3 volts and a 0 bit with 0 volts, we get the following table:

input	output
0	1
1	0

NOT Gate

We call these tables that show the input/output relationship of a digital circuit *truth tables*.

In a circuit diagram that consists of multiple gates, we typically represent each gate with a simple symbol that indicates its type. The symbol for a NOT gate is

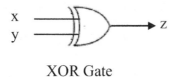

XOR, OR, NOR, AND, and NAND Gates

In this section we investigate several more gates. Like the NOT gate, these gates can be implemented using MOS transistors.

An *XOR* (exclusive OR) *gate* is a digital circuit that has two inputs and one output. It is a difference-detecting gate. That is, it outputs 1 if the values on its two input lines differ (i.e., one is 0 and the other is 1), and 0 otherwise. Here is the symbol that represents an XOR gate:

XOR Gate

The following table specifies the input-output relationship of an XOR gate:

x	y	z
0	0	0
0	1	1
1	0	1
1	1	0

The table shows that z is 1 if x and y have different values. Otherwise, z is 0.

One application of an XOR gate is to selectively complement (i.e., flip) a data bit. Consider the following configuration in which one input to an XOR gate is a control line, and the other input is a data line:

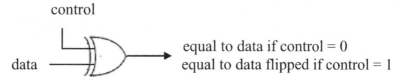

Suppose the control line is 0. If the data line is 0, then the two inputs are 0, in which case the gate outputs 0. If the data is 1, then the two inputs differ (0 on the control line and 1 and the data line), in which case, the gate outputs 1. Thus, if the control line is 0, the data appears unchanged on the output line. Now consider the action of the gate when the control line is 1. If the data is 0, then the two inputs differ, in which case the gate outputs 1. If the data is 1, then the two inputs are the same, in which case the gate outputs 0. Thus, if the control line is 1, the complement of the data appears on the output.

Chapter 4: Simple Digital Circuits

Control	Data	Output of XOR
0	0	0
0	1	1
1	0	1
1	1	0

- 0, 1: Data unaffected when control = 0
- 1, 0: Data flipped when control = 1

Rule: If the control line of an XOR is 0, the data passes through the gate unchanged. If the control line is 1, the data is complemented (i.e., flipped).

An *OR gate* is a digital circuit that has two or more inputs and one output. It is a not-zero-detecting gate. That is, it outputs 1 if its inputs are not all zero, and 0 otherwise. Here is the symbol that represents a two-input OR gate and its input/output relationship:

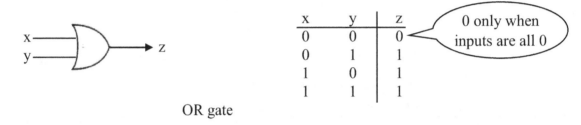

OR gate

x	y	z
0	0	0
0	1	1
1	0	1
1	1	1

0 only when inputs are all 0

A *NOR gate* is a digital circuit that has two or more inputs and one output. It is equivalent to an OR gate followed by a NOT gate ("NOR" is a contraction of "NOT OR"). A NOR gate is an all-zeros-detecting gate. That is, it outputs 1 if all its inputs are 0, and it outputs 0 otherwise. Here is the symbol that represents a two-input NOR gate and its input/output relationship:

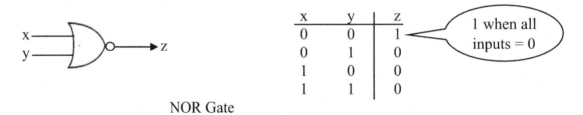

NOR Gate

x	y	z
0	0	1
0	1	0
1	0	0
1	1	0

1 when all inputs = 0

An *AND gate* is a digital circuit that has two or more inputs and one output. It is an all-ones-detecting gate. That is, it outputs 1 if all its inputs are 1, and 0 otherwise. Here is the symbol that represents a two-input AND gate and its input/output relationship:

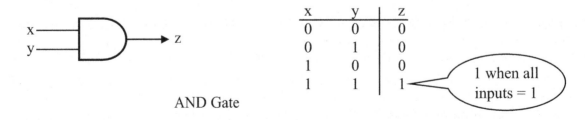

AND Gate

x	y	z
0	0	0
0	1	0
1	0	0
1	1	1

1 when all inputs = 1

One application of an AND gate is to selectively block or propagate data. Consider the following configuration in which one input to an AND gate is a control line, and the other input is a data line:

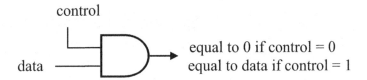

If the control input is 1, the data on the data input flows unchanged through the gate to the output. Think of the gate as being "open" so data can pass. If, however, the control is 0, then the gate is "closed" so data cannot pass through. Specifically, the output is held at 0 regardless of the data on the data input.

Control	Data	Output of AND
0	0	0
0	1	0
1	0	0
1	1	1

Rows with Control = 0: Gate is closed—output held at 0
Rows with Control = 1: Gate is open—data passes through

Rule: If the control line of an AND is 1, the data passes through gate unchanged. If the control line is 0, the output is held at 0.

A *NAND gate* is a digital circuit that has two or more inputs and one output. It is equivalent to an AND gate followed by a NOT gate ("NAND" is a contraction of "NOT AND"). It is a not-all-ones-detecting gate. That is, it outputs 1 if at least one of its inputs is 0, and 0 otherwise. Here is the symbol that represents a two-input NAND gate and its input/output relationship:

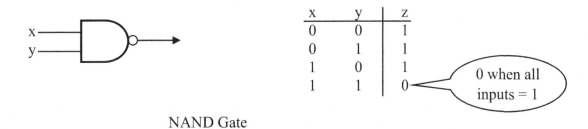

x	y	z
0	0	1
0	1	1
1	0	1
1	1	0

(0 when all inputs = 1)

NAND Gate

Tri-State Buffer

As the MOS diagrams in the preceding section indicate, the output of a NOT gate is either +3 volts (which represents the bit 1) or to 0 volts (which represents the bit 0). This is true of most gates. That is, their outputs can be in one of two states (3 volts or 0 volts). However, a *tri-state buffer* is an exception: Its output can be in any one of three states: 3 volts, 0 volts, or not connected to anything.

The term impedance refers to the opposition to electrical current flow. If a wire is not connected to anything, then there obviously is a high impedance between it and anything else. For this reason, we call the not-connected state that the output of a tri-state buffer can assume the *high-impedance* or *high-Z state* (Z is the abbreviation for "impedance"). Here is the symbol for a tri-state buffer and its truth table:

Input	Control	Output
0	0	High-Z
1	0	High-Z
0	1	0
1	1	1

Tri-State Buffer

When the control line is 1, the input passes unchanged through to the output. If, however, the control line is 0, then the output line is in the high-Z state. When the output line is in the high-Z state, it acts as if you had taken a scissors and cut the output line so it is no longer connected to anything.

Computers often use shared buses. A bus is a bundle of wires that connects one circuit in a computer to another. To avoid interference when one device is driving a shared bus (i.e., putting data on the bus), the other devices physically connected to the bus must be electrically disconnected from it. This is the job of tri-state buffers. For example, suppose two circuits are connected to the same one-wire bus. Circuit A uses the bus to send data to circuit 2. Circuit B uses the same bus to send data to circuit A. We call this kind a bus a *bidirectional bus* because data can flow in both directions. Only one circuit at a time can use the bus. Moreover, when one circuit is using the bus, the other circuit must be disconnected from the bus to avoid interference. Here is the circuit we need:

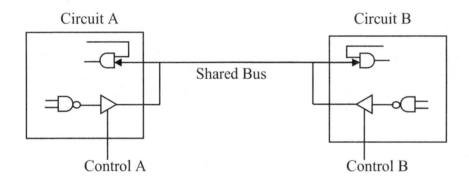

When circuit A needs to send data to circuit B, control A is set to 1 and control B to 0. Similarly, when circuit B needs to send data to circuit A, control B is set to 1 and control A to 0. There is no interference because when one circuit is using the bus, the NAND gate in the other circuit is disconnected from the bus by a tri-state buffer. If, however, the NAND gates were connected directly to the shared bus (i.e., no tri-state buffers), then each would attempt to drive the bus at the same time. For example, the NAND gate in circuit A might output a 1 while at the same time the NAND gate in circuit B might output a 0.

Here is an implementation of a tri-state buffer that uses a PMOS/NMOS pair:

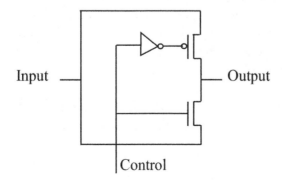

When the control input is 1, the two transistor switches are on. Thus, there is a direct connection from the input to the output. When the control input is 0, both switches are off, which isolates the output from the input. Here is the equivalent circuit when the control input is 0

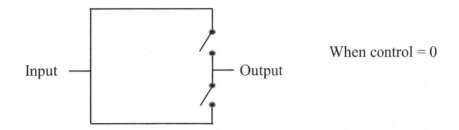

You can see that the output is not connected to anything.

The implementation of a tri-state buffer shown above uses CMOS (complementary MOS) technology. In CMOS technology, circuits consist of PMOS/NMOS pairs. Note that if we replace the PMOS transistor with an NMOS transistor in our implementation of a tri-state buffer, we can eliminate the NOT gate. But CMOS technology has some important advantages over using NMOS transistors exclusively. Thus, the CMOS implementation is superior.

Realizing Boolean Functions with Digital Logic

Truth tables represent Boolean functions. A Boolean function is a function whose inputs and outputs are all two-valued. Typically, the values of Boolean functions are either true/false or 1/0. Given any Boolean function in truth table form, we can easily implement a circuit consisting of AND, OR, and NOT gates that realizes that function.

Let's implement a circuit that realizes the XOR Boolean function:

x	y	z
0	0	0
0	1	1	← Select this row
1	0	1	← Select this row
1	1	0

To implement this function, we select only those rows whose output column contains a 1. Thus, for this function we select only the second and third rows. For each of these rows, we create a circuit using a single AND gate that will output 1 only when the inputs are as specified in that row. For example, for the second row we create the following circuit:

x ─▷∘─┐
      ├─AND─ Outputs 1 only if x = 0 and y = 1
y ────┘

This circuit will output a 1 if and only if its x input is 0 and its y input is 1 (the inputs for the second row). For the third row, we create a similar circuit:

x ────┐
      ├─AND─ Outputs 1 only if x = 1 and y = 0
y ─▷∘─┘

Finally, we feed the output of each of these circuits to a common OR gate and tie all the common inputs together to get

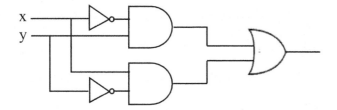

Let's analyze this composite circuit. If we input 0 and 1 for x and y, respectively, the top AND gate outputs a 1, causing the OR gate to also output a 1. Similarly, if we input a 1 and a 0 for x and y, respectively, then the bottom AND gate outputs a 1, again causing the OR gate to output a 1. However, all other input combinations cause both AND gates to output a 0, causing the OR gate to also output a 0. Thus, this circuit implements the truth table for the XOR function.

Since the XOR circuit is so important, we have a special symbol for it:

Half and Full Adders

A half adder is a digital circuit that add two bits. It outputs both a sum bit and a carry bit. Here are the four cases when adding two bits:

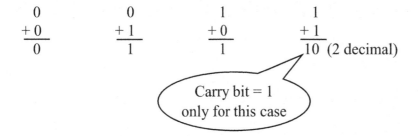

Note that the carry bit is 1 if and only both bits that are added are 1. Thus, the carry is equal to the AND of the two bits that are added. The sum bit is 1 if and only if the two bits that are added differ (i.e., 0 and 1 or 1 and 0). Thus, the sum bit is equal to the XOR of the two bits that are added.

Using our preceding observations on a half adder, we can easily implement it. We need an XOR gate (to compute the sum) and a AND gate (to compute the carry). Here is the circuit:

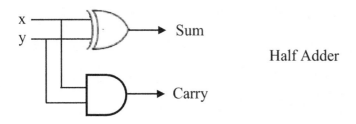

Half Adder

When we add two multi-bit binary numbers, for each column (except the rightmost column), we have to add three bits: the two bits in that column from the two binary numbers and the carry from the addition of the column to the right:

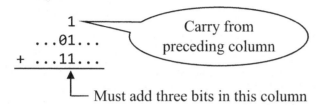

Must add three bits in this column

The circuit that adds three bits and outputs a sum bit and a carry bit is called a *full adder*. Here is the truth table for a full adder:

x	y	carry in	sum	carry out
0	0	0	0	0
0	0	1	1	0
0	1	0	1	0
0	1	1	0	1
1	0	0	1	0
1	0	1	0	1
1	1	0	0	1
1	1	1	1	1

In the preceding section, we learned how to implement a circuit that realizes a Boolean function. The resulting circuit consists of AND, NOT, and OR gates. We can do that for both the sum bit in a full adder and the carry out bit. However, a simpler way to implement a full adder is to use two half adders. The first half adder adds two bits and outputs a sum. Then the second half adder adds this sum and the third bit and outputs the three-bit sum. If either half adder produces a carry out, then the full adder should also produce a carry out. Thus, the full-adder carry out is the OR of the carry outs from the two half adders. Here is the circuit:

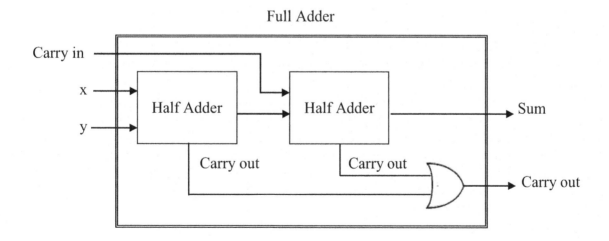

Sequential Circuits

A *sequential* circuit is a circuit that has memory. Because it has memory, its output can depend on the *sequence* of inputs leading up to and including the present input (hence the name "sequential").

Our first sequential circuit is the SR latch. An SR latch has two inputs labeled S and R It also has two outputs. These two outputs have values that are usually (but not always) the complements of each other (i.e., when one is 1, the other is 0). Accordingly, the outputs are labeled Q and Q' (Q' is the complement of Q). "S" stands for "set"; "R" stands for "reset". The SR latch consists of two NOR gates. The output of each NOR gate is fed back to the input of the other NOR gate:

SR Latch

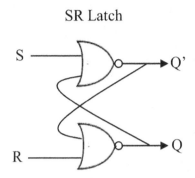

The truth table for an SR latch is unusual in that it includes *two* rows both corresponding to inputs S = R = 0:

S	R	Q	Q'
0	1	0	1
0	0	0	1
1	0	1	0
0	0	1	0
1	1	0	0

Two rows for S = R = 0

Let's examine the circuit and confirm that this table does, indeed, describe its operation. Let's start with the first row of the truth table (S = 0, R = 1). The 1 on the R input propagates through the bottom gate but is inverted before it exits. Thus, Q, the output of this gate is 0. The top gate has 0 on its S input and 0 (from the bottom gate) on its other input. The two 0's are ORed and inverted by the NOR gate. Thus, Q' is 1. A similar analysis can be done for the third row of the table. For the fifth row (S = R = 1), both gates necessarily output 0 because both have one input equal to 1.

Now here's the interesting part. Let's assume S = 0 and R = 1. Then the output of the top gate is 1. This 1 is fed back into the bottom gate, making both inputs to the bottom gate equal to 1. If we now change R from 1 to 0, the bottom gate still has one input equal to 1 (from the output of the top gate) which holds the output of the bottom gate at 0. The inputs of the upper gate do not change. Thus, its output also does not change. Changing R from 1 to 0 does not change the outputs of the circuit. The circuit is in a stable state (i.e., a state which will not change unless new inputs are applied). In this state, S = R = 0, Q = 0 and Q' = 1 (the second row of the table). We say the latch is *reset*.

Now let's start with the inputs S = 1 and R = 0. For this case, both inputs to the top gate are 1. If we now change S from 1 to 0, the top gate still has one input equal to 1 (from the output of the bottom gate) which holds the output of the top gate at 0. The inputs of the bottom gate do not change. Thus, its output also does not change. The circuit, therefore, is in another stable state (described by the fourth row of the table). In this state, S = R = 0, Q = 1, and Q' = 0. We say the latch is *set*. We have two possible outputs for the input S = R = 0, corresponding to the second and fourth rows of the truth table.

A *clock* in a computer is a circuit that outputs an alternating sequence of 1's and 0's. We often need a latch in a computer that can be set or reset only when a synchronizing signal (like the clock signal) is 1, and which can be set with a single data input. One type of latch with these properties is the clocked D latch:

Clocked D Latch

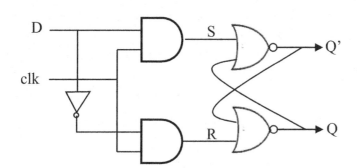

The back end of a clocked D latch is a SR latch. The two AND gates determine if the D data and the complement of D (provided by the NOT gate) get through to the SR latch. If clk = 1, then the AND gates are "open.". Thus, if clk = 1, the D data is passes through the upper AND gate to the S input of the SR latch; the complement of D passes through the lower AND gate to the R input of the SR latch. Thus, if D is 1, the SR latch is set. Similarly, clk = 1 and D = 0 the SR latch is reset. The set state of an SR latch represents the bit 1; the reset state represents the bit 0. Thus, when clk = 1 in a clocked D latch, its SR latch in effect stores the bit on the D input: if D = 1, the latch is set; if D is 0, the latch is reset. If, however, clk = 0, then the AND gates are "closed." Thus, the SR latch does not respond to the D input.

Typically, we use a clocked D latch in this way. To set the latch, we place a 1 on the D input, and then apply a 1 to the clk input. This combination provides the S = 1, R = 0 input to the internal SR latch, causing it to set. Similarly, to reset the latch, we place a 0 on the D input and apply a 1 to the clk input.

Here is the how we represent a clocked D latch in a circuit diagram:

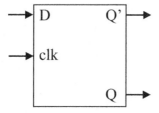

The clocked D latch sets or resets according to its D input *whenever* there is a 1 on its clk input. However, it is often the case that we need a one-bit storage device that sets or resets only during the brief period when the clk input is changing from 0 to 1 (or from 1 to 0). We call such a circuit a *flip-flop*.

The following is a graph of the clk signal versus time:

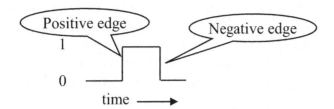

The edge of the graph corresponding to the transition from 0 to 1 is called the *positive edge*; the edge corresponding to the transition from 1 to 0 is called the *negative edge*. If a flop-flop sets or resets only during the positive edge of the clk signal, we say it is *positive edge-triggered*. If it sets or resets only during the negative edge, we say it is *negative edge-triggered*.

Here are the representations for positive and negative edge-triggered D flip-flops:

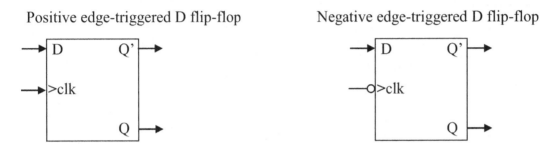

The greater-than sign on the clk input indicates the device is edge triggered. A bubble and a greater-than sign indicate the device is negative edge-triggered.

Two other flip-flops that are important are the SR flip-flop, which is the edge-triggered version of the SR latch, and the JK flip-flop. A JK flip-flop functions just like the SR flip-flop (the J input acts like the S input in an SR flip-flop, and the K input acts like the R input) except when J = K = 1. For this input, the JK flip-flop changes state each time a clock pulse is applied to the clk input. That is, if the flip-flop is set, it resets; if the flip-flop is reset, it sets.

Problems

1) Using four MOS transistors, construct a NOR gate. *Hint*: Connect two PMOS in series (i.e., end to end) and two NMOS in parallel.

2) Using two AND gates, one OR gate, and one NOT gate, implement the following function:

x	y	output
0	0	0
0	1	0
1	0	1
1	1	1

3) Implement the function in problem 2 using no gates.

4) Using AND, OR, and NOT gates, implement the following function:

x	y	z	output
0	0	0	0
0	0	1	0
0	1	0	0
0	1	1	0
1	0	0	0
1	0	1	1
1	1	0	1
1	1	1	1

5) What happens if the output of a NOT gate is fed back to its input?

6) What happens if the output of a tri-state buffer is fed back to its input when the control input is 1?

7) Suppose the four gates in a SR latch are replaced with NAND gates. Construct the truth table for the resulting latch. For which inputs are two states possible?

8) Construct a circuit that outputs 1 if and only if its three inputs are all 0. Use only two-input gates.

9) Construct a circuit that outputs 1 if and only if its four inputs are all 1. Use only two-input gates.

10) Construct the circuit that corresponds to the truth table for a full adder. Use an AND gate for each row in the truth table with a 1 output.

11) The *number of levels* in a circuit is the maximum number of simple gates in a path from an input to an output. Why is the number of levels important? What is the number of levels in the XOR gate? In a half adder implemented with an XOR gate and an AND gate? In a full adder implemented with two half adders?

12) What is the difference in function between an OR gate and an XOR gate? Why is an XOR gate called an exclusive OR gate.

13) What happens when an SR latch goes from the $S = R = 1$ state to the $S = R = 0$ state?

14) Modify the circuit for a clocked D latch so that it has two additional inputs: *clear* and *preset*. Both are *active high* (i.e., they have an effect only when set to 1). The clear input resets the latch regardless of the clk input. Similarly, the preset input sets the latch regardless of the clk input.

15) Same as problem 14 but for the latch in problem 7. The clear and preset inputs should be *active low*. That is, they have an effect only when set to 0.

16) Construct a two-bit register using two positive edge-triggered flip-flops. The clk signal to the register should cause the two flip-flops to be loaded from two data lines.

17) Construct a NOT gate using NAND gates exclusively.

18) Construct an AND gate using NAND gates exclusively.

19) Construct an OR gate using NAND gates exclusively.

20) Why is the NAND gate called a *universal gate*? *Hint*: See problems 17, 18, and 19.

5 Complex Digital Circuits

Introduction

In this chapter, we investigate complex digital circuits whose component parts are gates and flip-flops. For most of them, we will not investigate how they are implemented. However, you may find that you can easily implement them on your own using only what you learned about digital circuits from chapter 4.

16-bit AND, OR, XOR, and NOT Circuits

A *16-bit AND circuit* performs a bitwise AND operation on two 16-bit operands. A bitwise AND operation ANDs corresponding bits in the two operands. For example, consider the following bitwise AND operation:

```
  1010101010101010        operand 1
  1100110011001100        operand 2
  1000100010001000        result of bitwise AND
```

The two operand bits in each column are ANDed, each producing a one-bit result. If the two operand bits in a column are both 1, the result for that column is 1. Otherwise, it is 0.

A 16-bit AND circuit uses 16 AND gates. The 16 bits of one operand are applied to the left inputs of the 16 AND gates; the 16 bits of the other operand are applied to the right inputs of the 16 AND gates. Here is the representation for a 16-bit AND circuit:

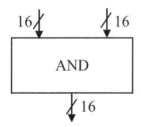

The cross-hatch labeled with 16 on each input indicates that each input is a 16-line bus (a *bus* is a bundle of wires that carries multi-bit data).

A *16-bit OR circuit* and a *16-bit XOR circuit* are configured just like an AND circuit but with 16 OR gates and 16 XOR gates, respectively, in place of the 16 AND gates. A *16-bit NOT* circuit is even simpler. It consists of 16 NOT gates, each one between an input line and its corresponding output line. Like the AND circuit, the OR, XOR, and NOT circuits perform bitwise operations.

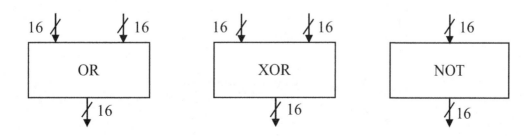

SEXT Circuit

A *SEXT* (sign extension) *circuit* has two inputs:

- a 16-bit operand with a right-justified signed number
- a 16-bit mask that indicates (with 1 bits) which bits in the first operand are occupied by the signed number. Suppose the signed number in the first operand occupies bit positions 0 to i. Then the mask should have 1 bits in positions 0 to i, and 0 bits in positions $i+1$ to 15.

For example, suppose the first operand is 1010101010101011 and its three rightmost bits (011) is the signed number it contains. Then the mask operand should have its three rightmost bits equal to 1 and the other bits equal to 0. Thus, the correct mask for this first operand is 0000000000000111. The output of the SEXT circuit is the signed number in the first operand sign-extended to 16 bits. In our example, the signed number in the first operand is 011. Because its sign bit is 0, it is extended to 16 bits with 0's. Thus, the SEXT circuit outputs 0000000000000011.

Now suppose the first operand to the SEXT circuit is again 1010101010101011 but its signed number is in its four rightmost bits. Then the correct mask is 0000000000001111. Because the rightmost *four* bits of the mask are 1's, the signed number in the first operand is its *four* rightmost bits: 1011. Because its sign bit is 1, it is extended to 16 bits with 1's. Thus, for this case the SEXT circuit outputs 1111111111111011. Here is the representation of the SEXT circuit:

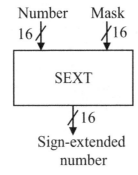

Multiplexer

A *multiplexer* is a combinational circuit that has multiple data inputs and one output. It also has control inputs. The control inputs determine which one of the multiple data inputs drives the output. In diagrams, multiplexers are usually represented with a trapezoidal figure. Here is the representation of a multiplexer with four data inputs:

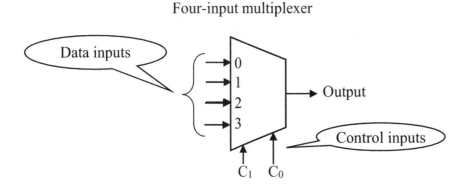

In this multiplexer, the two control inputs determine which data input drives the output. For example, if the control inputs are 10 (i.e., $C_1 = 1$ and $C_0 = 0$), which is 2 in decimal, then data input 2 is selected for output. Because this multiplexer has four data inputs (0, 1, 2, and 3), it needs two control inputs to specify the numbers 0 to 3. A multiplexer with eight data inputs needs three control inputs to specify the numbers 0 to 7 (000 to 111 in binary). In general, a multiplexor with 2^n data input needs n control lines.

In the multiplexer above, the data inputs are single lines. However, multiplexors can also have inputs which are buses. For example, here is a two-bus multiplexer:

Two-Bus Multiplexer

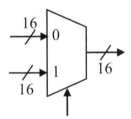

Note that the control input has no cross hatch. Thus, it is a single wire. If it is 0, then bus 0 drives the output bus; if it is 1, then bus 1 drives the output bus.

Decoder

1100 is the binary encoding of the number 12 decimal. A decoder "decodes" an unsigned binary number. That is, it is a circuit that indicates the number represented when given the binary encoding of that number. Here is the representation of a two-bit decoder:

Two-bit decoder

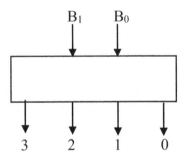

A two-bit binary number in applied to the B_1, B_0 inputs. The decoder responds by placing a 1 on the output line that corresponds to the inputted number. For example, suppose $B_1 = 1$ and $B_0 = 0$. Thus, 10 binary (2 decimal) is the inputted number. The decoder responds by placing a 1 on output line 2. It puts 0 on the three other lines.

Two-bit unsigned binary numbers can range from 0 to 3. Thus, a two-bit decoder has four output lines. An n-bit unsigned binary number can range from 0 to $2^n - 1$. Thus, an n-bit decoder has 2^n output lines.

Multi-Bit Addition and Subtraction

When we add two binary numbers by hand, we work from the rightmost column to the leftmost column. In each column except the rightmost, we have to add three bits: the two bits in that column plus a possible carry in from the column to the right. In each column, we have to determine two results: the sum of the bits added for that column and the carry out into the next column to the left. A computer adds binary numbers in exactly the same way. It has a sub-circuit—called a *full adder*—for each column that performs the computation for that column. Thus, an adder that adds two 16-bit numbers has 16 full adders.

A full adder has three inputs: carry in, top bit in, and bottom bit in. It has two outputs: sum out and carry out. If the addition of the bits on the three input lines produces a carry, the full adder outputs a 1 on the carry out line. Otherwise, it outputs a 0.

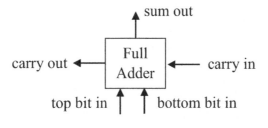

An *adder/subtractor circuit* consists of a full adder for each column to be added or subtracted, connected together in a serial fashion. For example, the following diagram shows the configuration of a three-bit adder/subtractor.

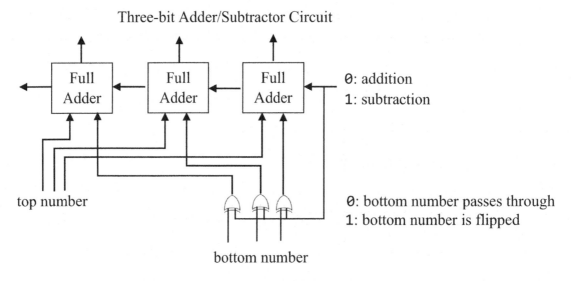

Three-bit Adder/Subtractor Circuit

This one circuit does both addition and subtraction. A 16-bit version of this circuit is part of the arithmetic-logic unit in the CPU of the LCC. We do not need separate circuits for addition and subtraction. On an addition, there is never a carry into the rightmost column. Thus, the carry-in line of the rightmost full adder in the adder circuit above is set to 0 on an addition. This input is also applied to the XOR gates so the bottom number passes unchanged into the full adders. On a subtraction, the carry-in line of the rightmost full adder is set to 1. This input is also applied to the XOR gates through which the bottom number passes, causing the bottom number to be flipped. The bottom number flipped plus the 1 applied to the carry-in input of the rightmost full adder *yields the two's complement of the bottom number*. Thus, the circuit in this case adds the two's complement of the bottom number to the top number, which is equivalent to subtracting the bottom number from the top number. Let's summarize what happens on a subtraction: The circuit above computes the result of the following subtraction

```
    N
  - M
```

where N and M are two's complement numbers by performing the following addition:

```
    N
   ~M    bottom number with its bits flipped
  + 1    carry into the rightmost full adder
```

For example, to subtract 2 from 5 (assuming four-bit numbers), the computer adds 5, 2 with its bits flipped, and 1:

```
  0101 = +5
  1101 = +2  with its bits flipped
     1
  ────
  0011 = 3
```

Note that the computer does *not* first takes the two's complement of the bottom number and then add it to the top number. This approach would require *two* add operations: one to add 1 to get the two's complement of the bottom number, and a second to add the complemented bottom number to the top number. Instead, the computer complements the bottom number and adds it to the top number *all in one operation*. For example, to subtract 0 for 0, the computer adds 0, 0 with its bits flipped, and 1 in one operation:

```
  0000 = 0
  1111 = 0  with its bits flipped
     1
  ────
  0000 = 0
```

Note that this computation produces a carry out of the leftmost column.

Signed Overflow

In a computer, the result of an addition or subtraction is usually stored in an area of fixed size. Thus, the range of values that can be stored is limited. For example, if the result of an addition or subtraction is stored in a 16-bit register, then the range of signed numbers that can be accommodated is −32768 to +32767 (a *register* is a storage area within the central processing unit of a computer). If the result of an addition or a subtraction is outside this range, we say *overflow* has occurred. If two positive numbers are added, the sum may be too big to fit into a fixed size storage area. Similarly, if two negative numbers are added, the result may be too negative to fit.

When a positive number and a negative number are added, overflow never occurs. The sum has to be less than the positive number because a negative number is added to the positive number. The sum also has to be greater than the negative number because a positive number is added to the negative number. Thus, the sum must lie between the negative number and the positive number. For example, suppose P is a positive number and N is a negative number. Then their sum lies between N and P:

58   Chapter 5: Complex Digital Circuits

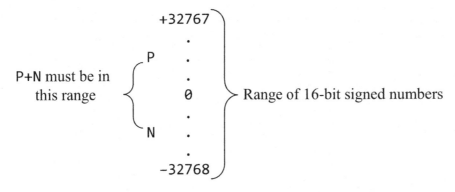

The sum cannot be more positive than the positive number or more negative that the negative number. Thus, overflow cannot occur.

A computer subtracts by adding the two's complement of the bottom number. Thus, the computer subtracts two numbers with the same sign by adding two numbers with different signs, in which case overflow cannot occur.

Rule: If two signed numbers *with different signs* are added or two signed numbers *with the same sign* are subtracted, overflow cannot occur.

If two signed numbers are added and overflow occurs, the sign of the result does not match the sign of the numbers added. Here are the two possible scenarios for overflow:

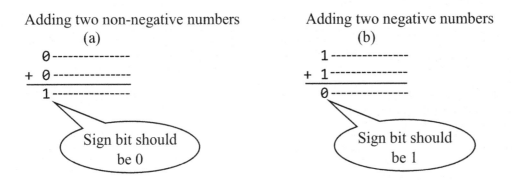

Here are all the possible scenarios for no overflow:

For case (a), there must have been a carry into the leftmost position because the bit in the result in that position is 1, but no carry out of the leftmost position. In case (b), there must have been no carry into the leftmost position because the result in that position is 0 (if there were a carry in, then the result bit would be 1). But there is a carry out of the leftmost position. Thus, for both overflow scenarios *the carry into the leftmost position does not match the carry out*. However, in all the non-overflow cases, the carry into the leftmost position matches the carry out. That is, if a carry in occurs then a carry out also occurs; if a carry in does not occur, then a carry out also does not occur. In case (c), neither a carry in nor a carry out of the leftmost position occurs. In case (d), both a carry in and a carry out occur. In case (e), both a carry in and a carry out occur. In case (f), neither a carry in nor a carry out occur.

Because the symptom of signed overflow is a mismatch between the carry into the leftmost position and the carry out, the computer hardware can easily test for signed overflow using a single XOR gate (recall that an XOR gate is a difference-detecting gate). In a computer, the full adder circuit that operates on the leftmost bits has a carry-in line and a carry-out line. A 1 on these lines indicates a carry; a 0 indicates no carry. Thus, if the values on the carry-in and carry-out lines differ, then signed overflow has occurred. If the carry-in and carry-out lines are applied to an XOR gate, its output is 1 if its input values differ (in which case overflow has occurred) or 0 if the input values are the same (in which case overflow has not occurred):

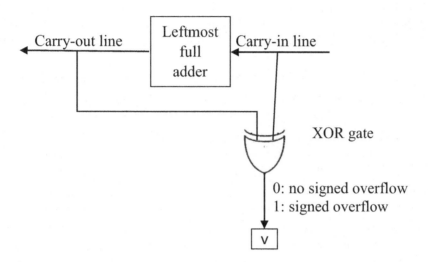

The output of the XOR gate is saved in a one-bit register named v (for oVerflow). Thus, if v is 1 after an addition or subtraction of signed numbers, then overflow has occurred. v is called a *flag* because it flags a condition that results from an arithmetic-logic unit operation.

Rule: In an addition or subtraction of signed numbers, overflow has occurred if the carry into the leftmost position does not match the carry out.

Rule: If overflow occurs during the addition of two signed numbers, the sign bit of the computed result is wrong. If the sign bit of the computed result is 0, then the sign bit of the *true* result is 1 (thus, the true result is negative). If the sign bit of the computed result is 1, then the sign bit of the true result is 0 (thus, the true result is greater than or equal to zero). The same is true for a subtraction of signed numbers.

Unsigned Overflow

The symptom of overflow when two *unsigned numbers* are added is a *carry out of the leftmost position*. This carry indicates another bit is needed to hold the result:

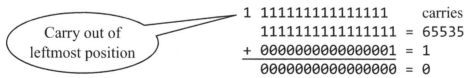

On a subtraction of unsigned numbers, if the bottom number is larger than the top number, the result is negative. But the smallest unsigned number is 0. Thus, any negative result on a subtraction indicates unsigned overflow. If we subtract two unsigned numbers using the *borrow technique* (i.e., the subtraction technique you learned in grade school), a borrow into the leftmost position indicates that the top number is smaller than the bottom number. Thus, a *borrow into* the leftmost position indicates unsigned overflow on a subtraction. But a computer *does not subtract using the borrow technique*—it subtracts by adding the two's complement of the bottom number to the top number. It turns out that in a subtraction, if a *borrow in occurs* with the borrow technique, then a carry out *will not occur* with the two's complement technique. If a *borrow in does not occur* with the borrow technique, then a carry out *will occur* with the two's complement technique. Thus, in a subtraction of unsigned numbers, the computer hardware can determine if overflow has occurred from the carry out of the leftmost position: 1 indicates a borrow in would *not* have occurred if the borrow technique were used—thus, no overflow; 0 indicates a borrow in would have occurred if the borrow technique were used—thus, overflow. In an addition of unsigned numbers, *the test for overflow is the reverse*: on an addition, a carry out of 1 indicates signed overflow; a carry out of 0 indicates no signed overflow.

On an addition, the carry out of the leftmost position is stored in a one-bit register called the c flag. To make the tests for signed and unsigned overflow the same, the LCC *on a subtraction* (but not on an addition) flips the carry out with an XOR gate before storing it in the c flag. This modification of the carry out bit makes the overflow tests for the addition and subtraction of unsigned numbers the same: 1 in the c flag indicates unsigned overflow; 0 indicates no unsigned overflow.

In the circuit that follows, on a subtraction, the bits of the bottom number are flipped and a 1 is applied to the carry-in line of the rightmost full adder (which has the effect of two's complementing the bottom number). The 1 applied to the carry in of the rightmost full adder is also applied to the *b* input of the XOR gate that outputs to the c flag. Thus, on a subtraction (but not an addition), the carry out of the leftmost position is flipped by the XOR gate before it is stored in the c flag. Thus, for this circuit, on a subtraction, the c flag acts like a borrow flag. On a subtraction, 1 in the c flag indicates that a borrow into the leftmost position would have occurred if the subtraction were performed with the borrow technique. For this reason, the c flag is sometimes called the c/b flag (c for "carry", b for "borrow").

In addition to the v and c flags, the LCC also has an n flag and a z flag. The n flag is set to the leftmost bit of the result in an addition or subtraction (see the following diagram). Thus, the n flag is set to 1 if the result is negative (because the leftmost bit of a negative value is 1), and to 0 otherwise. The z flag is set to the output of a NOR gate whose inputs are all the bits in the result of an addition or a subtraction. Recall that a NOR gate is a zero-detecting gate. That is, if all its inputs are 0, it outputs 1. Otherwise, it outputs 0. Thus, the z flag is set to 1 if the result of an addition or subtraction is 0, and to 0 otherwise.

Rule: In an addition of unsigned numbers, a carry out of the leftmost position indicates overflow. In a subtraction, no carry out of the leftmost position (or a borrow in if the borrow technique is used) indicates overflow.

Three-bit Adder/Subtractor Circuit with n, z, c, and v Flags

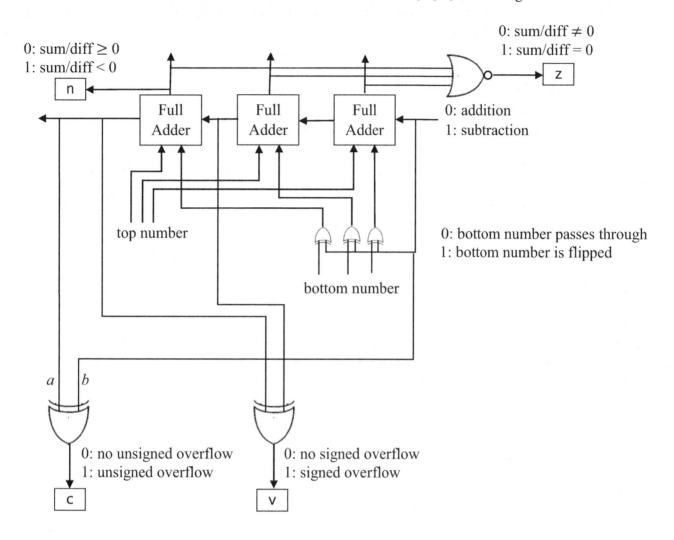

Barrel Shifter

Suppose a CPU register contains the following number (−7 decimal):

`1111111111111001`

If the register is *shifted right logically* two positions, all the bits in the register move two positions to the right. Two 0's are injected in the left of the register to occupy the positions vacated by the shift. The two rightmost bits are shifted out of the register. The register would then contain

`0011111111111110`

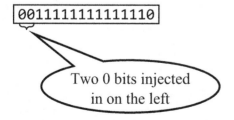

Two 0 bits injected in on the left

If the register is *shifted left logically*, the bits move to the left the specified number of positions and 0 bits are injected into the right side of the register. Thus, a shift left logical works the same way a shift right logical works except in the opposite direction.

A *shift right arithmetic* works like a shift right logical except copies of the sign bit—not 0's—are used to occupy the positions vacated by shift. For example, suppose the following register is shifted right arithmetically two positions:

`1111111111111001`

The register would then contain

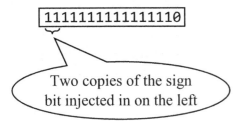

Another type of shift operation is a *rotate*. A rotate is analogous to the game of musical chairs: The bits shifted out of one side of a register and injected back into the other side of the register. For example, suppose the following register is rotated right two positions:

`1111111111111001`

The register would then contain

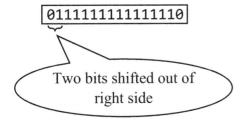

A *barrel shifter* is a combinational circuit that shifts a data word a specified number of positions. Here is the representation of a 16-bit barrel shifter that performs a logical right shift:

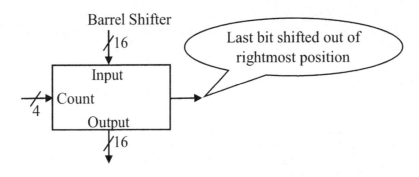

The data on the input lines are shifted right logically the number of positions specified by the count input. The shifted data appears on the output lines. The count input consists of four lines. Thus, it can specify a count in the range of 0000 to 1111 (0 to 15 decimal). The shifter also outputs the last bit shifted out of the rightmost position. This bit is routed to the c (carry) flag in the CPU.

To shift a register contents, the register's contents and the shift count have to be routed to the barrel shifter. The output of the barrel shifter then has to be routed back to the register.

We name the shift left logical, shift right logical, shift right arithmetic, rotate left, and rotate right operations sll, srl, sra, rol, and ror, respectively.

Multiplier and Divider/Remainder Circuits

The LCC has a multiplier circuit and a divider/remainder circuit (a remainder circuit provides the remainder that results when two integers are divided). These circuits have two 16-bit inputs (for the two numbers to be operated on) and a 16-bit output. Here are their representations:

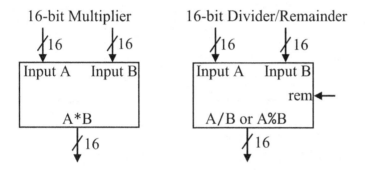

Random-Access and Read-Only Memory

Random-access memory (RAM) is memory that is both readable and writeable. It is "random" in the sense that any word can be read or written without having to access the words that precede it. Read-only memory (ROM) also has this random-access property. However, by convention, "random-access memory" refers only to readable/writeable memory.

RAM can be implemented using rows of D flip-flops, with one row for each word in memory. For example, we can implement the memory on the LCC with 65536 rows, each containing 16 D flip-flops. A decoder is used to determine which row is used in a read or write operation. Two control inputs, rd and wr, determine if a read or write operation occurs. Data is transferred between RAM and the memory data register (mdr) in the CPU via a bidirectional data bus. The address to read from (on a read operation) or the address to write to (on a write operation) is transferred from the memory address register (mar) in the CPU to RAM via a unidirectional address bus:

64   *Chapter 5: Complex Digital Circuits*

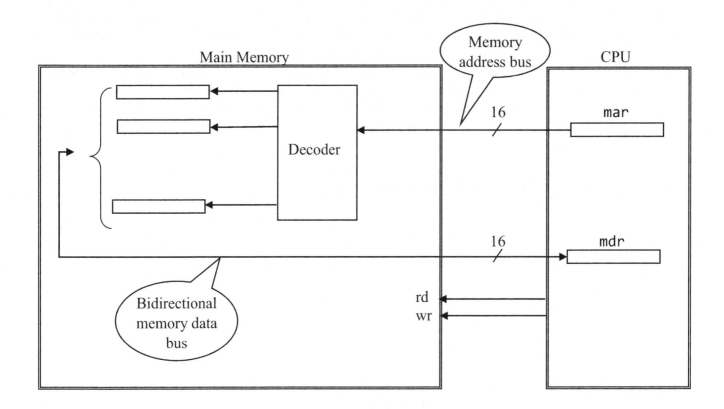

RAM is volatile. That is, its contents are lost when power is turned off. Read-only memory, on the other hand, is non-volatile. That is, its contents are maintained whether or not power is on. Moreover, its contents are permanent. That is, its contents cannot be modified.

In the LCC, a ROM is used for the microstore (recall that microstore is the memory that holds the microprogram). Its interface is simpler than that of a RAM: It does not have the rd and wr control inputs. In addition, it is not connected to a bidirectional data, but to a strictly outgoing data bus. An incoming bus provides the address, and the outgoing data bus the provides the microinstruction at that address:

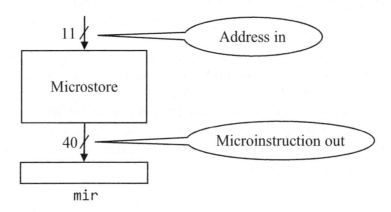

The ROM that contains the microprogram has 2048 locations whose addresses therefore, run from 0 to 2047. Eleven-bit unsigned numbers range from 0 to 2047. Thus, we need 11 bits to address the locations in microstore.

Registers

A register is a special memory area that holds one item of information. The CPU of the LCC has 32 16-bit registers named r0, r1, ..., r31. Here is the representation of r0 and its associated circuitry:

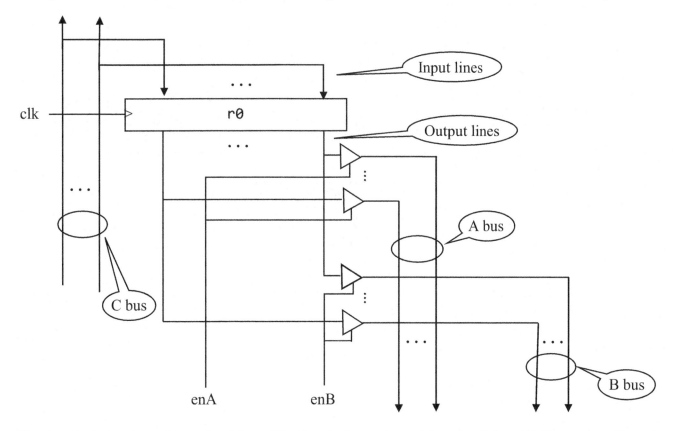

If enA = 1, the register drives the A bus. If enB = 1, the register drives the B bus. If clk goes positive, the register is loaded from the C bus.

When we represent a register in a circuit diagram, we will include in its representation the tri-state circuitry. For example, here is our representation of r0:

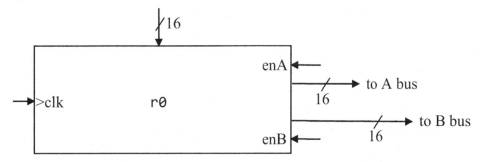

Arithmetic-Logic Unit

One typically first hears about an ALU (arithmetic-logic unit) in an introductory computer science course. We learn that an ALU is the unit within a computer that performs computations at incredible speeds. Surely, the ALU, the heart of a computer, must be awesomely complex. But this, in fact, is not necessarily the case. The ALU for LCC is actually quite simple, consisting of a straightforward combination of mostly simple circuits.

Let's examine the operation of the ALU in the LCC from an external point of view:

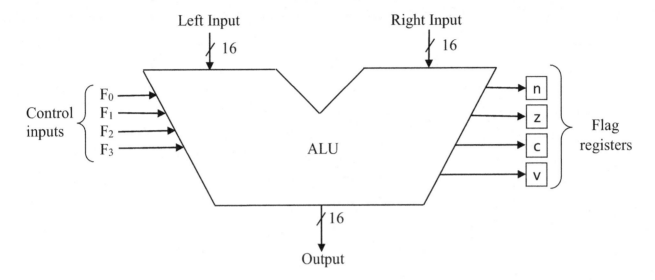

The ALU has two 16-bit input ports (left and right) and a 16-bit output port. Four control inputs, F_0, F_1, F_2, and F_3, determine the operation performed by the ALU. These four control inputs can specify a number from 0000 to 1111 (0 to 15 decimal), each of which triggers a different ALU operation as indicated by the following table:

F_3	F_2	F_1	F_0		Operation	Output	Flags Set
0	0	0	0	(0)	nop	left	
0	0	0	1	(1)	not	~left	nz
0	0	1	0	(2)	and	left & right	nz
0	0	1	1	(3)	sext	left sign-extended, (right is mask)	nz
0	1	0	0	(4)	add	left + right	nzcv
0	1	0	1	(5)	sub	left − right	nzcv
0	1	1	0	(6)	mult	left * right	nz
0	1	1	1	(7)	div	left / right	nz
1	0	0	0	(8)	rem	left % right	nz
1	0	0	1	(9)	or	left | right	nz
1	0	1	0	(10)	xor	left ^ right	nz
1	0	1	1	(11)	sll	left << right (logical)	nzc
1	1	0	0	(12)	srl	left >> right (logical)	nzc
1	1	0	1	(13)	sra	left >> right (arithmetic)	nzc
1	1	1	0	(14)	rol	left << right (rotate)	nzc
1	1	1	1	(15)	ror	left >> right (rotate)	nzc

The first two operations in this table operate on the left input only. For these operations, the ALU ignores the right input. The first operation (when $F_0 = F_1 = F_2 = F_3 = 0$) is a no-operation. That is, it allows the data from the left input to pass unchanged through the ALU. For this operation, the flag registers are not set.

It is easy to construct the ALU given the circuits for each operation it is to perform. We simply use a bus multiplexer that selects which circuit determines the ALU's output:

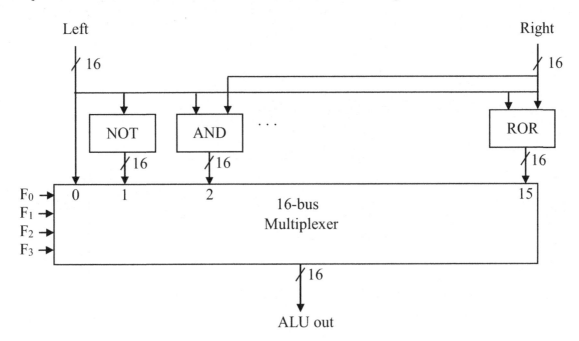

Loadable Binary Counter

A binary counter is a register that can hold a binary number. However, unlike a regular register, each time it receives an external clock pulse, the number it holds increases by one. When it contains all ones and then receives a clock pulse, it wraps around to all zeros.

The *microprogram program counter* (mpc) in the LCC is a binary counter. The function of the mpc at the microlevel is like the function of the pc register at the machine level. The pc register at the machine level holds the main memory address of the machine instruction to be executed next. Similarly, the mpc register at the microlevel holds the microstore address of the next microinstruction to be executed.

Because microstore has only 2048 slots, to address it requires only an 11-bit address (11-bit unsigned numbers range from 0 to 2047). For this reason, the mpc is an 11-bit counter. The address in the mpc is inputted to microstore. Microstore responds by outputting the microinstruction at that address to the microinstruction register (mir). As soon as the microinstruction is loaded into the mir, its bits are outputted to the various control inputs of the computational circuits in the LCC. For example, four of the output lines on the mir go to the F_0, F_1, F_2, and F_3 control inputs of the ALU. We refer to the process of loading a microinstruction into the mir and letting its bits drive the various control inputs in the CPU as *executing the microinstruction*.

Microstore has 2048 slots each of which is 40 bits wide. For this reason, in the diagram below, we label it with "2048 x 40."

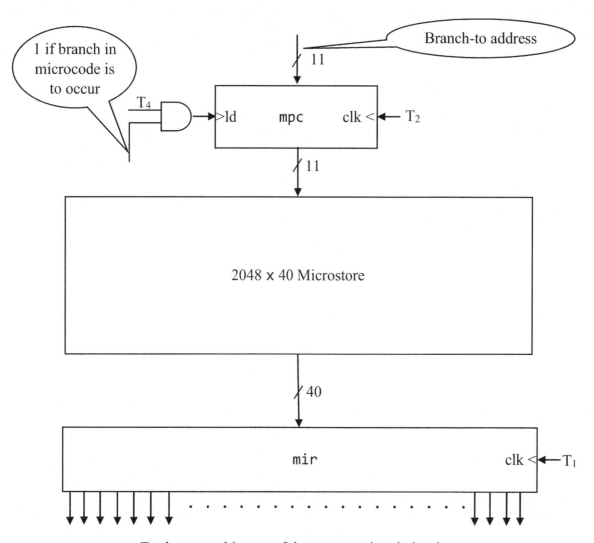

To the control inputs of the computational circuits

To execute the next microinstruction, a clock plus is applied to the clk input of the `mpc`, causing it to be incremented. The new address in the `mpc` causes microstore to output the next microinstruction to the `mir`. Thus, this mechanism causes microinstructions in microstore to be executed in memory order. For example, after the microinstruction at address 0 is executed, the microinstruction at address 1 is executed, then at address 2, and so on. However, occasionally a branch in microcode has to occur. For example, suppose the `mpc` currently contains 5, but the next microinstruction to be executed is at the address 9, not 6. To effect a branch to location 9, the branch-to address—9—is applied to the 11 input lines on the `mpc`, and a 1 is applied to its ld input. This causes the branch-to address to be loaded into the `mpc`, which in turn causes the microinstruction at that address to be outputted by microstore and loaded into the `mir`. The result is that the microinstruction at location 9 is executed next.

Clock Sequencer

The various operations that a CPU performs often require a specific sequence of suboperations. For example, to add 1 to the `pc` register, the CPU must

1) Route the `pc` register contents and the number 1 to the adder circuit in the ALU.
2) Instruct the ALU to add.
3) Route the sum back to the `pc` register.

Obviously, the CPU cannot route the sum back to the `pc` register before the ALU has produced the sum. The ALU cannot produce the sum until it has the `pc` register contents and the number 1. Thus, the three steps above must be performed in the order given.

To enforce a particular order of operations to be performed, the LCC uses a sequencer circuit that outputs four clock signals: T_1, T_2, T_3 and T_4. When T_1 is 1, T_2, T_3, and T_4 are 0. When T_1 returns to 0, T_2 becomes 1. When T_2 returns to 0, T_3 becomes 1. When T_3 returns to 0, T_4 becomes 1. When T_4 returns to 0, the cycle repeats with T_1 again becoming 1. Here is a timing graph that shows the T_1, T_2, T_3, and T_4 signals versus time:

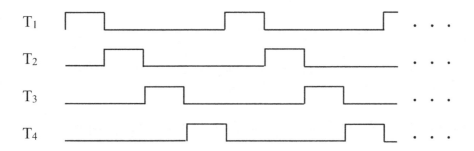

T_1 drives the clock input on the `mir`. Thus, every time T_1 goes to 1, a new microinstruction is loaded into the `mir`. T_2 is applied to the clk input of the `mpc`, causing the `mpc` to be incremented at T_2. The microinstruction at the new address immediately appears on the output lines of microstore. However, this microinstruction is not immediately loaded into the `mir` because the `mir` is loaded from microstore only at T_1.

If the two inputs to the AND gate that is driving the ld input on the `mpc` are 1, the AND gate outputs a 1 to the ld input, which causes the `mpc` is loaded with the branch-to address. Because one of the inputs to this AND gate is T_4, the loading of the branch-to address into the `mpc` occurs only at T_4 and only if the other input to the AND gate is 1. Then on the next T_1, the next microinstruction—either at the incremented address (if no branch) or at the branch-to address—is loaded into the `mir` and executed.

Let's summarize: A microinstruction is executed simply by loading it into the `mir`. Once loaded into the `mir`, the bits in the microinstruction immediately flow out of the `mir` to the control inputs of the computational circuits in the CPU via wires that connect the bits in the `mir` to these control inputs. The computational circuits respond by performing those operations specified by the microinstruction.

At T_1, the `mir` is loaded from microstore with the microinstruction at the address in the `mpc`. At T_2, the `mpc` is incremented. At T_3 the flag registers are updated (unless the ALU is performing a no-operation). At T_4, if a branch in microcode is to occur, the branch-to address is loaded into `mpc`, replacing the address there, and a register is loaded from the C bus (unless the specified register is the read-only register `r31`). Every T_1-T_2-T_3-T_4 cycle, one microinstruction is executed.

Microinstructions are executed in memory order unless a branch occurs. The microinstructions determine what operations are performed. The T_1, T_2, T_3, and T_4 clock signals determine the order in which suboperations occur. In the next chapter, we will see what determines if a branch in microcode occurs and where the branch-to address comes from.

Because the `mpc`, microstore, and the `mir` together provide the control signals for the CPU, they are collectively called the *control unit*.

The operations that are triggered by a microinstruction and the clock signals are called *micro-operations*. The execution of one machine language instruction requires the execution of a sequence of micro-operations. For example, consider the following add machine instruction in the basic instruction set: 2007. When executed, it adds the word at address 007 in main memory to the ac register. It requires the following sequence of micro-operations:

1. Send the address 007 to main memory.
2. Initiate a main memory read operation.
3. Route the operand read from main memory and the contents of the ac register to the left and right inputs of the ALU, respectively.
4. Instruct the ALU to add its left and right inputs.
5. Route the output of the ALU to the ac register.

Problems

1) The mir is loaded at T_1, the mpc is incremented at T_2, and the mpc is loaded at T_4 with the branch-to address if a branch is to occur. Why is this order necessary?

2) Are four clock subcycles necessary? Would three (T_1, T_2, and T_3) be sufficient? Explain.

3) At what point in the T_1-T_2-T_3-T_4 clock cycle does the ALU start performing its calculation?

4) Can the clock input to the mpc be driven by T_3 instead of T_2? Explain.

5) The standard registers in the LCC can drive either the A bus or the B bus. Can a register drive both buses simultaneously?

6) Implement a two-input multiplexer using AND, OR, and NOT gates.

7) Implement a two-input multiplexer using tri-state buffers and one NOT gate.

8) Implement a two-bus multiplexer in which each bus has three wires.

9) Implement a four-bit bitwise AND circuit.

10) Implement a four-bit bitwise NOT circuit.

11) Implement a two-bit decoder.

12) What items have to be provided to main memory for a read operation.

13) What items have to be provided to main memory for a write operation.

14) The LCC has a bank of registers consisting of 31 standard read/write registers and one read-only register. How many control inputs does the entire bank of registers have?

15) There are more control inputs in the LCC than there are bits in a microinstruction. So how can a single microinstruction drive all the control inputs? *Hint*: A decoder has n inputs and 2^n outputs.

16) How do the pc and mpc registers differ? Are they implemented in the same way? Are they incremented in the same way? Are they the same size? To where are there outputs directed?

17) If microstore had 8192 slots, what would be the size of the mpc register?

18) Using the schematic representations of the sub-circuits that make up the ALU of the LCC, implement the ALU. In particular, show how the add, subtract, divide, and remainder operations are activated. Recall that the adder/subtractor circuit has a control input that determines which operation it performs. In your implementation show what drives this control input. Do the same for the divider/remainder circuit.

19) Design a two-input decoder that has an additional control input X. When X = 1, the decoder should work normally. However, when X = 0, the decoder outputs should all be forced to 0.

20) Design a 2-bit ALU that has one control input C. When C = 0, the output should be the NAND of its inputs; when C = 1, the output should be the AND of its inputs. Use as few gates as possible.

21) Design a circuit with 8 data inputs and one output P. P should be set equal to 0 if the parity of the data bits is even, and to 1, otherwise. Parity is even if the number of 1 bits is even; parity is odd if the number of 1 bits is odd.

22) When the adder/subtractor subtracts a number from itself, does a carry out of the leftmost position always occur? Consider 0 subtracted from 0, and 5 subtracted from 5.

23) A 64-word RAM needs a decoder with 64 outputs, one for each word. Show how the decoder with 64 outputs can be replaced by two eight-output decoders, or three four-output decoders. What is the advantage of the latter two approaches?

24) Construct a two-bit counter. *Hint*: Use JK flip-flops.

25) Give the truth table for a full subtractor circuit. A full subtractor has two data inputs, a borrow-in, a difference-out, and a borrow-out.

6 Microlevel of the LCC

Introduction

In this chapter, we construct the LCC using the digital circuits we studied in the preceding two chapters. Our LCC, however, will still not be a working computer. The one item that will be lacking is the microcode that implements the basic instruction set that we learned in chapters 2 and 3. In the next chapter, however, we will learn how to write the microcode for the basic instruction set and incorporate it into the LCC.

Data Path

The *data path* of the LCC consists of a bank of 32 16-bit registers (named r0 to r31), the ALU, the A and B buses that connect the register bank to the left and right inputs of the ALU, and a C bus that connects the output of the ALU to the register bank:

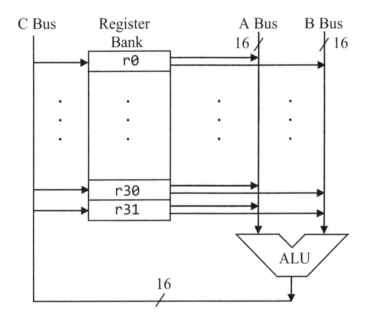

Registers r0 through r30 are read/write registers. Register r31 is read-only—it permanently contains 0. Each register can drive the A and/or B buses through a bank of tri-state buffers. Each register, except the read-only register r31, can be loaded from the C bus.

The data path in the LCC forms a circuit. Data in the registers flows through the A and B buses into the ALU. The result computed by the ALU then flows through the C bus back into one of the registers.

One of the operations that the ALU can perform (when $F_0 = F_1 = F_2 = F_3 = 0$) is to simply let its left input pass unchanged to its output. This operation is performed by the ALU when a microinstruction is executed that loads one register from another. For example, to load r2 from r1, a microinstruction is executed that makes the data in r1 take the circular route from r1 straight through the ALU to r2.

Main Memory Interface

The data path forms a circuit. For it to do any useful work, data from main memory has to get into the circuit. Then when the computed results are available, the results have to get out of the circuit back into main memory. This data flow between the data path and main memory is accomplished with two buses (the *memory data bus* and the *memory address bus*), two registers in the register bank (the mar and the mdr), and two control signals (rd and wr) that emanate from the microinstruction in the mir:

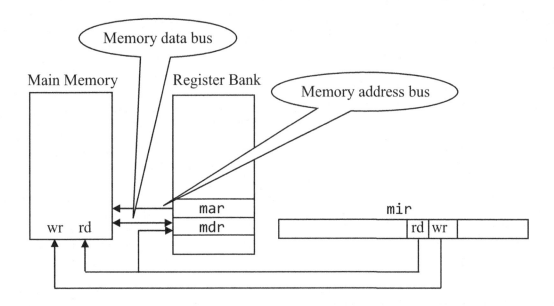

To read from main memory, a microinstruction first has to load the mar (the memory address register), which is r29 in the register bank, with the address to read from. When the read-from address is in the mar, a second microinstruction has to be executed with 1 in its rd field. This 1 drives the rd input on the main memory unit, causing it to read the word at the given address and provide it to the mdr via the memory data bus. The rd signal from the microinstruction is also applied to the circuitry surrounding the mdr. The rd signal causes the mdr to be loaded from the memory data bus.

To write to main memory, microinstructions first have to be executed that load the mar with the write-to address and the mdr, which is r30 in the register bank, with the data to be written. When both the mar and mdr have been properly initialized, another microinstruction has to be executed with 1 in its wr field. This 1 then drives the wr input on the main memory unit, causing it to write the word in the mdr to the address given by the mar.

Decoding the Register Fields in a Microinstruction

Recall that each register in the register bank is connected to the A bus via a bank of tri-state buffers. The enA (enable A) control line for a register is connected to the control inputs of all the tri-state buffers that connect that register to the A Bus. Thus, if 1 is on the enA control line for a register, that register drives the A bus. That is, the tri-state buffers let the contents of the register pass through to the A bus. If, on the other hand, the enA control input is 0, then the tri-state buffers electrically disconnect the register from the A bus. If you are fuzzy at this point on how registers work, you should review the section on registers in the preceding chapter.

Each register has an enA control input. Thus, there are a total of 32 enA control inputs. The A field in a microinstruction specifies which register should drive the A bus. But the A field has only five bits. How can five bits correctly drive all 32 enA control inputs? The answer is via a decoder. The A field in a microinstruction contains the 5-bit number of the register that is to drive the A bus. This number is inputted to a decoder. A five-input decoder has 32 outputs, one for each enA control input:

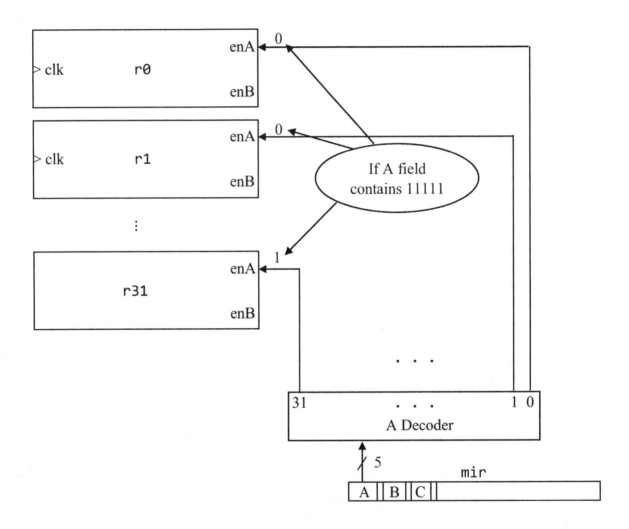

For example, suppose the A field in the microinstruction in the `mir` contains 11111 (31 decimal). Then the decoder outputs 1 on its output 31 and 0 on all its other output lines. Thus, only `r31` drives the A bus. Note that to make the diagram easy to read, we have omitted in the diagram above the connections between the registers and the A and B buses.

The B field similarly drives a B decoder whose outputs drives the enB control inputs of the registers. Thus, the B field in the microinstruction determines which register drives the B bus.

The C field drives a C decoder whose outputs drive AND gates that in turn drive the clk inputs on registers `r0` through `r30`. For example, here is the configuration at the clk input for `r30`:

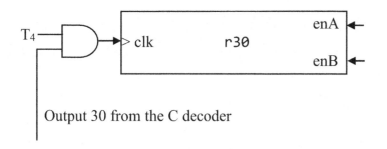

Output 30 from the C decoder

The second input to the AND gates that drive the clk inputs is T_4. Thus, if a register is to be loaded, it is loaded only at T_4. r31 is a read-only register so output 31 of the C decoder is not used. The C field determines which register is loaded from the C bus.

Our description here of the circuitry that performs register decoding is somewhat simplified. For now, it is adequate for our purposes. In chapter 9, we will see the actual decoding circuitry in the LCC.

Specifying the ALU Operation

The ALU field of the microinstruction in the mir determines the ALU operation. The ALU field is a four-bit field. A four-bit field can hold unsigned numbers from 0000 to 1111 (0 to 15 decimal). Thus, the ALU field can specify any one of 16 operations.

Here is a summary of the operations that the ALU can perform:

F_3 F_2 F_1 F_0	Mnemonic	Output	Flags Set
0 0 0 0 (0)	nop	left	
0 0 0 1 (1)	not	~left	nz
0 0 1 0 (2)	and	left & right	nz
0 0 1 1 (3)	sext	left sign-extended, (right is mask)	nz
0 1 0 0 (4)	add	left + right	nzcv
0 1 0 1 (5)	sub	left − right	nzcv
0 1 1 0 (6)	mult	left * right	nz
0 1 1 1 (7)	div	left / right	nz
1 0 0 0 (8)	rem	left % right	nz
1 0 0 1 (9)	or	left | right	nz
1 0 1 0 (10)	xor	left ^ right	nz
1 0 1 1 (11)	sll	left << right (logical)	nzc
1 1 0 0 (12)	srl	left >> right (logical)	nzc
1 1 0 1 (13)	sra	left >> right (arithmetic)	nzc
1 1 1 0 (14)	rol	left << right (rotate)	nzc
1 1 1 1 (15)	ror	left >> right (rotate)	nzc

Branch-Control Logic

As we previously discussed, after the mpc is incremented at T_2, a new address can be loaded into the mpc at T_4, triggering a branch in the microcode. This new address comes from the addr field of the current microinstruction (i.e., the microinstruction in the mir):

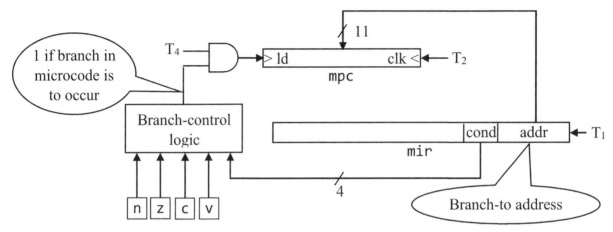

The output of the branch-control logic drives one of the inputs of the AND gate on the mpc. Thus, if the branch-control logic outputs a 1, then at T_4, a 1 is applied to the ld input of the mpc, causing the branch-to address to be loaded into the mpc. This, in turn, causes a branch in microcode to the branch-to address. The output of the branch-control logic depends on eight inputs: the four flag registers (n, z, c, and v) and the four-bit cond field in the microinstruction in the mir. If the cond field is 0000, then the branch-control logics output a 0 regardless of the contents of the flag registers, in which case a branch does not occur. If the cond field is 1111 (15 decimal), then the branch-control logic outputs a 1 regardless of the flag registers, in which case a branch occurs. The other 14 cond field values cause a branch only if the flag registers have specific values. For example, if the cond field is 0100 (4 decimal), then the branch-control logic outputs 1 if the n flag register contains 0 (which indicates the result of the most recent ALU operation was not negative). Thus, the cond field value 0100 in a microinstruction indicates that the microinstruction is a "branch on not negative" instruction. Similarly, the cond field value 0010 in a microinstruction indicates that the microinstruction is a "branch on not zero" instruction.

Here is list of all the cond field values, what they test, and their effect:

cond	Mnemonic	Branch if	Branch on
0	nobr		never
1	zer	z = 1	zero
2	!zer	z = 0	not zero
3	neg	n = 1	negative
4	!neg	n = 0	not negative
5	cy or <	c = 1	less than (unsigned compare/overflow)
6	!cy or >=	c = 0	greater than or eq (unsigned compare)
7	v	v = 1	signed overflow
8	pos	n = z	positive
9	lt	n != v	less than (signed compare)
10	le	n != v or z = 1	less than or equal (signed compare)
11	gt	n = v and z = 0	greater (signed compare)
12	ge	n = v	greater than or equal (signed compare)

13	<=	c = 1 or z = 1	less than or equal (unsigned compare)
14	>	c = 0 and z = 0	greater than (unsigned compare)
15	br		always

In the next chapter, the only cond values we use in the implementation of the basic instruction set are 0 (no branch), 2 (branch on not zero), 4 (branch on not negative), and 15 (unconditional branch). We will discuss the other cond values in later chapters when we need to use them.

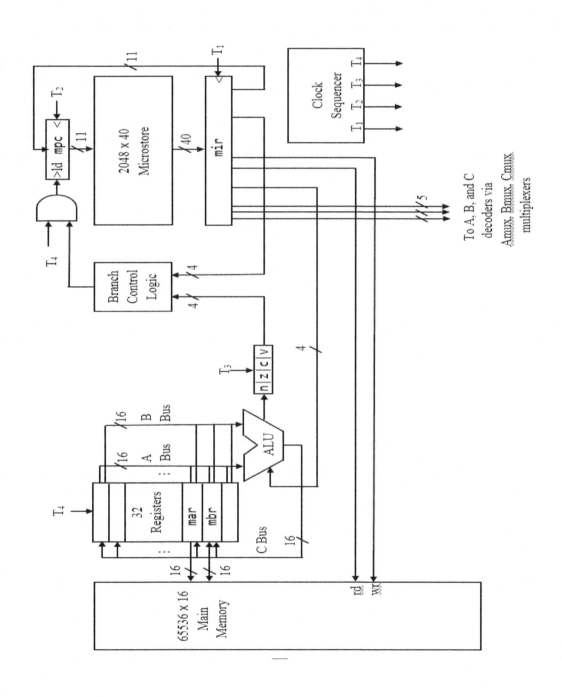

Complete Microlevel of the LCC

Problems

1) Implement the clock sequencer circuit. Use three D flip-flops. Assume that the left flip-flop is preset with 1 and the middle and right flip-flops are preset with 0. *Hint*: Drive each D input with the Q' output of the preceding flip-flop. Assume you have available a clock circuit that outputs an alternating sequence of 1's and 0's, each 1 and 0 lasting 1 microsecond. Drive you clock sequencer with the output of the clock circuit.

2) Would there be any advantage if the mar were a register separate from the register bank? From where would it be loaded? What would determine if it is to be loaded with a new address?

3) Design the circuitry associated with the mdr. Recall that the mdr drives the bidirectional memory data bus and can be loaded from either the C bus (if rd = 0) or from the memory data bus (if rd = 1). *Hint*: You will need tri-state buffers and a multiplexer.

4) Assume the cond field in the microinstruction contained only two bits that specified the following four actions: no branch (00), branch on negative (01), branch on zero (10), and unconditional branch (11). Design the branch-control logic for such a cond field.

5) If the n flag is not equal to the v flag after a subtraction, how do the numbers used in the subtraction compare?

6) If the c flag is equal to 1 after a subtraction, how do the numbers used in the subtraction compare?

7) If the n flag equals the z flag after a subtraction, is the result of the subtraction negative, zero, or positive?

8) How many more bits would a microinstruction have to have if the A, B, and C decoders were not used.

9) Why does the ALU have to have a no-operation function (when $F_0 = F_1 = F_2 = F_3 = 0$)?

10) Is it possible to do a bitwise OR operation without using the OR operation in the ALU?

11) With the LCC configured as described, is it possible for two registers to drive the A bus at the same time? Would you ever want two registers to drive the A bus at the same time?

12) With the LCC configured as described, is it possible for two registers to be loaded from the C bus at the same time? Would you ever want two registers to be loaded from the C bus at the same time?

13) Why is the rd signal applied to the mdr?

14) Which circuit or circuits in the LCC most likely limit how short the clock cycles an be?

15) Implement a SEXT circuit that inputs a five-bit signed number and outputs its eight-bit equivalent.

7 Microprogramming the Basic Instruction Set

Introduction

In this chapter, we create the microcode that implements the basic instruction set that we learned in chapters 2 and 3. You should print out a copy of the file `sim.txt` that is in the software package for this book. This file provides all the information you will need to create and use microcode on the LCC. It is a reference that you will frequently want to refer to when you are writing microcode.

Format of a Microinstruction

The following list shows each field of a microinstruction along with its width in bits:

A	Amux	B	Bmux	C	Cmux	ALU	u	rd	wr	cond	addr	
5	1	5	1	5	1	4	1	1	1	4	11	width

To implement the basic instruction set, we do not need to use all the fields in a microinstruction. In this chapter we will study only those fields we need for the basic instruction set (A, B, C, alu, rd, wr, cond, and addr). We will study the remaining fields (Amux, Bmux, Cmux, and u) when we implement a more complex instruction set (the optimal instruction set) that requires them.

Alternate Register Names

The 32 registers in the register bank in the LCC are named "`r0`", "`r1`", ..., "`r31`". Most, however, have alternate names that are easy to remember (and, therefore, are more convenient to use). For example, "`ac`" is the alternate name for the `r0`; "`pc`" is the alternate name for `r28`. Some of the registers have specific initial values. These registers are given alternate names that indicate their contents. For example, the alternate name for `r31` is "`0`". This register contains the constant 0. The names "`1`", "`3`", and "`4`" are alternate names for the registers that contain the corresponding constants.

The alternate name for `r21` is "`m8`". This name indicates the register contains the constant consisting of all 0's except for its eight rightmost bits, which contain all 1's. Thus, in hex it contains 00ff. Similarly, the register whose alternate name is "`m12`" contains the constant 0fff (four 0 bits followed by twelve 1 bits).

Some of the registers contain a 16-bit value that has only a single 1 bit. The alternate names for these registers consist of the prefix "`bit`" followed by the position of the 1 bit. For example, the register whose alternate name is "`bit5`" contains a 1 bit in position 5 and 0 bits elsewhere. Thus, in hex it contains 0020 (recall that bits are numbered right to left, starting with 0).

When we write microcode, we sometimes need a register to hold some value temporarily. For this purpose, we use `r11`, whose alternate name is "`temp`".

Here is a listing of the registers with their alternate names and initial contents:

Register Number	Name	Initial Contents	Function
0	r0 or ac		accumulator register
1	r1		
2	r2		
3	r3		
4	r4		
5	r5 or fp		frame pointer register
6	r6 or sp		stack pointer register
7	r7 or lr		link register
8	r8		
9	r9		
10	r10		
11	r11 or temp		
12	r12 or 3	0x0003	
13	r13 or 4	0x0004	
14	r14 or bit5	0x0020	
15	r15 or bit11	0x0800	
16	r16 or bit15	0x8000	
17	r17 or m3	0x0007	
18	r18 or m4	0x000f	
19	r19 or m5	0x001f	
20	r20 or m6	0x003f	
21	r21 or m8	0x00ff	
22	r22 or m9	0x01ff	
23	r23 or m11	0x07ff	
24	r24 or m12	0x0fff	
25	r25 or ir		machine instruction register
26	r26 or dc		decoding register
27	r27 or 1	0x0001	constant 1
28	r28 or pc		program counter register
29	r29 or mar		memory address register
30	r30 or mdr		memory data register
31	r31 or 0	0x0000	read-only register containing 0

In microcode, if we want to access the constant 1, we simply specify the register that contains the constant 1, which conveniently has the alternate name "1". When can similarly access the constants 0, 3, and 4 using the names "0", "3", and "4", respectively. But note that there is no register with the name "2" that contains the constant 2. Thus, in microcode we cannot access the constant 2 simply by specifying the name "2". To get 2, we have to use a microinstruction that adds 1 (which is available in a register) to 1. Be sure to understand that the register bank contains only a very small subset of constants. These are the only constants which we can access directly.

Symbolic Microcode

A microinstruction on the LCC consists of 40 bits. We can write microcode directly in binary. But doing that would be a very tedious and error-prone process. A much better way to write microcode is to use an easy-to-use symbolic form and then use an assembler to translate the symbolic form to the required 40-bit binary form. In other words, we can do at the microlevel exactly what we did at the machine level: Write code in symbolic form and then assemble it to the required binary form.

Let's look at a microinstruction in binary form that increments the `pc` register by 1. To increment the `pc`, we need a microinstruction that will add the current contents of the `pc` register and the constant 1, and store the sum back into the `pc` register. Here is the microinstruction that will do this (the unspecified fields are all 0's):

```
11100  11011  11100  0100
  A      B      C    ALU
```

The A field of the microinstruction contains 11100 (28 decimal), which is the number of the `pc` register. The B field contains 11011 (27 decimal), which is the number of the register than contains the constant 1. Thus, the contents of the `pc` register are placed on the A bus, and the constant 1 is placed on the B bus. The ALU field contains 0100 (4 decimal) which specifies the add operation. Thus, the `pc` register contents and 1 are added, and the sum is outputted to the C bus. The C field, like the A field, contains the register number of the `pc` register. Thus, the sum is loaded into the `pc` register.

The order in which these operations occur is controlled by the clock sequencer. At the start of T_1, this microinstruction is loaded into the `mir`. During T_1, the A and B decoders enable the `pc` register and the register with 1 to drive the A and B buses. During T_2, the ALU performs the addition operation. At T_3, the flag registers are set. Finally, at the start of T_4, the `pc` register is loaded with the incremented value.

Now let's write this microinstruction in symbolic form:

```
a.pc add b.1 c.pc        ; increment pc register
```

The `add` component of this instruction indicates that the ALU should perform an add operation (so the ALU field in the microinstruction should be 0100). The component `a.pc` indicates that the A field should contain the number of the `pc` register The component `b.1` indicates that the B field should contain the register number of the register whose name is "1". `c.pc` indicates that the C field should contain the number of the `pc` register. This instruction does not specify values for the other fields of the microinstruction. For that reason, those fields default to all 0's. For example, the binary microinstruction will have 0000 in its cond field. Thus, a branch will not occur when this microinstruction is executed. The rd and wr fields are also 0 so neither a read nor a write operation will occur. Note that a comment in symbolic microcode starts with a semicolon.

To specify the ALU operation in a symbolic microinstruction, we use the mnemonic for the operation. For example, we use `add` for addition, `and` for bitwise AND, and `sext` for sign extension. See the file `sim.txt` for a complete list of the ALU mnemonics.

To specify a read or write operation, we use the mnemonics `rd` or `wr`, respectively. For example, here is the two-instruction sequence of microinstructions that fetches the instruction in main memory that the `pc` register points to and increments the `pc`:

```
a.pc c.mar               ; load mar with contents of pc
a.pc add b.1 c.pc rd     ; read from address in mar and incr pc
```

In the first instruction above, we have not specified an ALU operation. Thus, the ALU field defaults to 0000, which causes the ALU to let its left input pass through to its output unchanged. Since this instruction puts the pc register contents on the A bus which is connected to the left side of the ALU, the contents of the pc register pass unchanged to the output of the ALU, and then to the C bus where it is loaded into the mar.

Before a read operation can be performed, the mar has to be initialized with the address to read from. Thus, the rd operation in the sequence above cannot be in the first microinstruction (the microinstruction that loads the mar). However, a rd can occur simultaneously with the add operation. So the rd operation can appear in the second microinstruction. We, of course, could perform the same operations with three microinstructions:

```
a.pc c.mar              ; load mar with contents of pc
rd                      ; read from address in mar
a.pc add b.1 c.pc       ; increment pc
```

But this sequence is inefficient: It requires more time and occupies more space in microstore.

The fields of a microinstruction all default to all 0's if the symbolic microinstruction does not specify a value for them, *with one exception*. If the C field is not specified, it defaults to 11111 (31 decimal), which is the register number of the read-only register that contains 0. Because it is read only, it is not loaded with a new value when the C field contains its default value. Thus, if a symbolic microinstruction does not specify a value for the C field, then no register in the register bank is loaded with a new value. For example, the only effect of the second microinstruction in the three-instruction sequence above is to read from memory—no register in the register bank is loaded with a new value.

To specify a branch in a microinstruction, we specify the mnemonic for the branch condition, "@", and the label on the instruction to branch to (for a list of the branch-condition mnemonics, see the file sim.txt). For example, the following instruction branches to the label fetch if the output of the ALU (which is equal to the current contents of the ac register) is zero:

```
a.ac add b.0 zer@fetch              ; branch to fetch if ac = 0
```

zer is the mnemonic for branch on zero. It is translated to 0001 (branch on zero) and placed in the cond file of the microinstruction. The address corresponding to the label fetch is placed in the addr field of the microinstruction. The output of the ALU is equal to the contents of the ac register plus 0 (because the add operation adds the contents of the ac register on the A bus with 0 on the B bus). We need the add operation here to force the ALU to set the flag registers.

The ALU operation occurs at T_2. At the start of T_3, the flag registers are loaded based on the current ALU output. Thus, the if ac register contains zero, then the ALU output is zero, and the z flag register accordingly is set to 1. Otherwise, it is set to 0. During the rest of T_3, the branch-control logic generates an output that depends on the flag registers and the cond field in the microinstruction. Because the cond field in this instruction specifies a branch on zero, the branch-control logic will output 1 if z = 1 and 0 otherwise. Then at T_4, if the branch-control logic outputs a 1 (which means z = 1, which means the ac register contains 0), the branch-to address (the address in the addr field of the microinstruction that corresponds to the label fetch) is loaded into the mpc, causing a branch to that location in the microcode. Thus, this microinstruction branches to fetch if the ac register contents are 0. If, on the other hand, the ac register contents are nonzero, then the microinstruction in the next higher location in microstore is executed next.

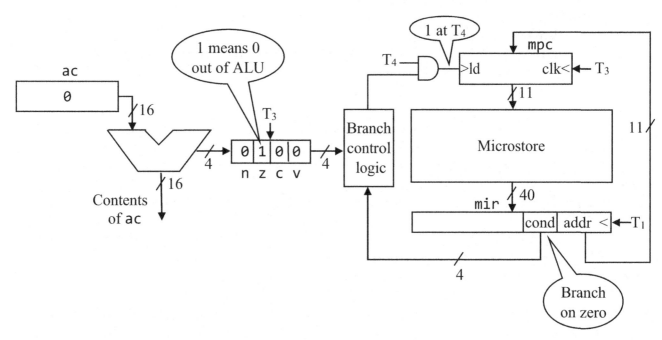

The components of a symbolic microinstruction can be in any order except for a label. If a symbolic microinstruction has a label, the label should be the first component and start in column 1. All the components of a symbol microinstruction, except for a label, should not start in column 1. For example, here is an instruction with the label `fetch`:

```
fetch    a.pc c.mar
```

Fetching a Machine Instruction and Decoding the Opcode

Recall that the CPU performs the following four steps repeatedly:

1. Fetch machine instruction that `pc` register points to.
2. Increment the `pc` register
3. Decode the opcode.
4. Execute the instruction just fetched.

Here is the microcode that performs steps 1 and 2 (with line numbers added):

```
1 fetch     a.pc c.mar
2           a.pc add b.1 c.pc rd
```

The instruction on line 1 causes the `pc` register to drive the A bus and the `mar` to be loaded from the C bus. Because an ALU operation is not specified, the default operation is in effect: namely, whatever enters the left side of the ALU appears unchanged on its output. The A bus drives the left side of the ALU, and the ALU output drives the C bus. Thus, whatever is placed on the A bus (which for this instruction is the contents of the `pc` register) appears unchanged on the C bus. Because the instruction on line 1 loads the `mar` from the C bus, its net effect is to load the `mar` from the `pc` register.

Line 1 loads the `mar` with the desired address. Line 2 then performs the read operation. Specifically, it reads the instruction into the `mdr` from the address provided by the `mar`. Recall that we first have to load the mar with the desired address. Then perform the read. Thus, line 1 cannot both load the mar and perform the read. Line 2 also increments the pc register.

Now that we have the machine instruction in the mdr, the next step is to decode its opcode. The opcode is in the machine instruction's leftmost four bits. Its first bit resides in the sign-bit position. Thus, to determine the first bit of an opcode, we can simply use a microinstruction that passes the instruction through the ALU so that the n flag register reflects the sign bit. We can then use a microinstruction that branches if n = 1 or falls through to the next instruction if n = 0. Then to determine each of the remaining bits in the opcode, we use microinstructions that shift the machine instruction left one position (so that the next opcode bit occupies the sign-bit position) and repeat the test given above. We call this process *decoding the opcode*.

Let's examine the microinstructions that determine the first bit of the opcode:

```
3           a.mdr add b.0 c.ir neg@L1      ; add 0 to set flag registers
4 L0        a.ir sll b.1 c.dc neg@L01      ; shift left 1 position
5 L00       a.dc sll b.1 c.dc neg@L001     ; shift left 1 position
```

Line 3 copies the contents of the mdr (which contains the machine instruction) via the ALU to the ir. Thus, the n flag register reflects the leftmost bit of the machine instruction. The add in this microinstruction *is necessary*. If omitted, the default ALU operation is in effect in which case the ALU will not set the flag registers. The conditional branch on line 3 branches to L1 if the n flag is 1 (which means the first bit of the opcode is 1), or it falls through to the next instruction if the n flag is 0 (which means the first bit of the opcode is 0). The labels we use reflect what the opcode must be at that label. For example, if we reach the microinstruction at the label L00, the first two bits of the opcode must be 00—hence, the label L00.

Line 4 loads the dc (decoding) register with the machine instruction now in the ir. But the ALU performs a sll operation as the machine instruction passes through the ALU. The B bus provides the shift count. Line 4 puts 1 on the B bus, which results in a shift of one position. Thus, the leftmost bit in the output of the ALU is now the second bit of the opcode, and the dc register is loaded with the shifted instruction. The conditional branch instruction on line 4 tests the n flag, which now reflects the second bit of the opcode. It branches to L01 if the second bit is 1, or it falls through to the next instruction whose label is L00. Thus, if we reach the label L01, then the first two opcode bits are 01; if we reach the label L00, the first two opcode bits are 00.

This process of shifting the machine instruction and then testing its leftmost bit is continued until all four bits of the opcode have been determined. Each opcode causes the decoding process to end up at the label for that opcode. For example, if the opcode is 0000, then the decoding process ends up at the label L0000; if the opcode is 0001, then the decoding process ends up at the label L0001, and so on.

The shifting occurs in the dc register—not the ir register. To execute the machine instruction, the microprogram needs the original unshifted machine instruction. Using the dc register for shifts leaves the original machine instruction unchanged in the ir.

Here is the microcode for the entire fetch, increment pc, and decoding processes (the microcode that interprets the machine instructions is not included):

```
           ;================================================
           ; Fetch machine instruction, increment pc
fetch      a.pc c.mar
           a.pc add b.1 c.pc rd


           ;================================================
           ; Decode instruction
           a.mdr add b.0 c.ir neg@L1
L0         a.ir sll b.1 c.dc neg@L01
L00        a.dc sll b.1 c.dc neg@L001
L000       a.dc sll b.1 c.dc neg@L0001
           br@L0000

L1         a.ir sll b.1 c.dc neg@L11
L10        a.dc sll b.1 c.dc neg@L101
L100       a.dc sll b.1 c.dc neg@L1001
           br@L1000

L01        a.dc sll b.1 c.dc neg@L011
L010       a.dc sll b.1 c.dc neg@L0101
           br@L0100

L11        a.dc sll b.1 c.dc neg@L111
L110       a.dc sll b.1 c.dc neg@L1101
           br@L1100

L001       a.dc sll b.1 c.dc neg@L0011
           br@L0010

L011       a.dc sll b.1 c.dc neg@L0111
           br@L0110

L101       a.dc sll b.1 c.dc neg@L1011
           br@L1010

L111       a.dc sll b.1 c.dc neg@L1111
           br@L1110


           ;================================================
           ; Interpret machine instruction
L0000      ; ld =============================================
              .
              .
              .
L1111      ; trap ===========================================
           br@fetch ; only microcode needed for opcode 1111
```

Interpreting Machine Instructions with Microcode

When the CPU is executing a machine language instruction, it is really *executing microinstructions*—specifically, the sequence of microinstructions that have the effect the machine instruction is supposed to have. We say that the microcode is *interpreting* the machine code.

To complete the microcode for the basic instruction set, we have to provide a sequence of microinstructions for each machine language instruction that interprets that machine instruction. For example, if the machine instruction is a ld instruction, its opcode is 0000. Thus, the decoding process ends up at the label L0000. At this label, we need the microcode that interprets the ld instruction.

The ld instruction has the address in its rightmost 12 bits from which it is to load. The address must be extracted from the instruction, zero-extended to 16 bits, and loaded into the mar. A read operation then reads the operand from main memory into the mdr. Finally, the contents of the mdr have to loaded into the ac register. Sounds complicated, but the microcode is actually fairly simple:

```
L0000         ; ld ==========================================
              a.ir and b.m12 c.mar   ; extract address from inst and zero-extend
              rd                     ; read operand from main memory into mdr
              a.mdr c.ac br@fetch    ; load ac from mdr and start again
```

The first instruction extracts the address in the machine instruction (which is in the ir) by ANDing it with m12. Recall that m12 is the name of the register that contains 0fff. The ANDing process in effect zero-extends the 12-bit address to 16 bits (because the leftmost four bits of m12 are 0000). This instruction then loads the extracted address into the mar, in preparation for the read operation that the next microinstruction performs. The read operation reads the operand at the specified address into the mdr. The third microinstruction then copies the mdr to the ac register, completing the interpretation of the ld instruction. The unconditional branch in the third instruction branches back to fetch, which starts the processing of the next machine instruction. These three microinstructions should be placed in the microcode given in the preceding section at the label L0000.

Let's look at the microcode for the str instruction. The str instruction stores the ac register contents at the address given by the contents of the sp register plus the relative address in the str instruction. For example,

```
    str 3
```

stores the ac register contents three words above the location sp points to whose address is the contents of the sp register plus 3. Here is the required microcode:

```
L0101         ; str =========================================
              a.ir and b.m12 c.mar   ; extract rel addr from str inst
              a.mar add b.sp c.mar   ; add the contents of sp to it
              a.ac c.mdr             ; load mdr from ac
              wr br@fetch            ; perform write and start next fetch
```

The call instruction is an interesting instruction because it involves a push operation (of the pc register contents). Here is its microcode:

```
L1010       ; call ==========================================
            a.sp sub b.1 c.sp          ; decrement sp
            a.sp c.mar                 ; load mar from sp
            a.pc c.mdr                 ; load mdr with return addr in pc
            wr                         ; push return addr onto stack
            a.ir and b.m12 c.pc br@fetch ; branch to subroutine
```

Recall that in a push operation, first the `sp` register is decremented, then the value to be pushed is stored in the location the `sp` register points to. The first microinstruction above decrements the `sp` register. The next two instructions prepare the `mar` and the `mdr` for the write operation in the fourth microinstruction that completes the push operation. The final instruction causes an unconditional branch to the address in the rightmost 12 bits of the `call` instruction. It does this by extracting the address from the `call` instruction using the AND operation and then loading it into the `pc` register.

The `call` instruction performs an unconditional branch. Here the microcode for the conditional branch instruction `brz`:

```
L1110       ; brz ===========================================
            a.ac add b.0 !zer@fetch      ; start fetch of next inst if ac != 0
            a.ir and b.m12 c.pc br@fetch ; branch to address in brz inst
```

In the first instruction, a register is not specified for the C field. Thus, the C field defaults to 31, which is the number of the read-only register. Thus, no register is loaded from the C bus. Nevertheless, the contents of the `ac` register still go through the ALU, causing the setting of the z flag register. The conditional branch in the first instruction branches back to `fetch` if z = 0, which is the case if the `ac` register contents are not zero (`!zer` is the mnemonic for branch on not zero). Thus, if the `ac` register is not zero, the branch occurs. Specifically, the machine instruction in main memory following the `brz` is fetched and executed next. If, however, the `ac` register is zero, then the branch in the first microinstruction instruction is not taken. Instead, second microinstruction is executed. It causes a branch to the address in the 12 rightmost bits of the `brz` instruction. Thus, a branch to the label given in the `brz` instruction occurs if the `ac` register contains 0.

A `trap` machine instruction (opcode 1111) is essentially a call of a service module in the operating system that performs an I/O operation or a `halt`. It is *not* implemented in microcode. The only microinstruction required for the trap instruction is an unconditional branch back to `fetch`:

```
L1111       ; trap ==========================================
            br@fetch ; only microcode needed for opcode 1111
```

Assembling and Using Microcode

Once you have completed writing the microcode for the basic instruction set, you can run programs written for this instruction set on the LCC simulator (see Problem 1 at the end of this chapter).

Suppose the file `b.sm` contains the complete microcode implementation of the basic instruction set (we use the extension ".sm" for symbolic microcode files). To assemble the microcode in `b.sm` to binary form, assemble it using the `micro` program by entering on the command line micro b.sm   (on Windows)

or

./micro b.sm   (on Linux, Mac OS X, or Raspbian)

The `micro` program responds by assembling the code in `b.sm` to binary and outputting the binary form to the file `b.m`. It also displays the files it uses:

```
Symbolic microcode file:  b.sm
Binary microcode file:    b.m
List file:                b.lst
```

The file `b.m` contains the translated microcode. The file `b.lst` is a text file that includes the source microcode and its translated form in hex.

Next, test your microcode by assembling the basic assembly language program in the file `btest.a` by entering

basic btest.a   (on Windows)

or

./basic btest.a   (on Linux, Mac OS X, or Raspbian)

The `basic` program translates the assembly language program in `btest.a` to binary and outputs the binary form to the file `btest.e`. Finally, run the machine language program in `btest.e` on the LCC simulator by entering

sim btest.e   (on Windows)

or

./sim btest.e   (on Linux, Mac OS X, or Raspbian)

You should see on the screen the following (which `sim` also writes the to the file `btest.log`):

```
Microlevel Simulator Version 1.0 Copyright (c) 2019 by Anthony J. Dos Reis
Opening machine code file btest.e
Opening microcode file b.m          ⎫ Using microcode for
Opening log file btest.log          ⎭ basic instruction set
========================================= output
12345678910
Correct if 1 to 10 displayed
=========================================
Machine code size:                  52 (dec)
Machine instructions executed:      47 (dec)
Microcode size:                     71 (dec)
Microinstructions executed:        276 (dec)
Load point:                          0 (hex)
```

If you do not see on your screen the numbers from 1 to 10, your microcode has a bug. Note that in the example above, `sim` is using the microcode for the basic instruction set (in the file `b.m`).

Debugging Microcode

The `sim` program has two features that are helpful for debugging microcode: the *step/trace function* and the *breakpoint instruction*. If you specify the argument "!" on the command line in addition to the name of the executable file when you invoke `sim`, the step/trace function is activated. For example, to run the executable file in `btest.e` with the step/trace function activated, enter on the command line

    sim btest.e !

`sim` will then execute and trace only one microinstruction each time you hit the Enter key. The trace includes the microinstruction itself and its effect. If instead of immediately hitting the Enter key, you first enter a *single* nonzero digit (1 to 9), then thereafter `sim` will execute that number of lines each time you hit the enter key. If you enter 0, the step function is canceled. If you enter q, the run is terminated. The trace that appears on the screen is also written to the ".log" file that `sim` creates.

The following is a sampling of the trace output that appears on the display screen when the program in `btest.e` runs. On the first prompt in this example, the Enter key is hit without first entering any characters, causing a trace of one microinstruction. On the next prompt, 2 is entered, causing the trace of two microinstructions before the next prompt (thereafter, hitting just the Enter key causes the trace of two microinstructions). Finally, q is entered, terminating the run.

```
      0: a.pc c.mar
mar = 0000/0000 nzcv = 0000

Hit Enter key to step, digit to change step, q to quit:
      1: a.pc add b.1 c.pc rd
pc = 0000/0001 read machine inst 0027 from loc 0 nzcv = 0000

Hit Enter key to step, digit to change step, q to quit: 2
      2: a.mdr add b.0 c.ir cond=neg addr=7
ir = 0000/0027 nzcv = 0000

      3: a.ir sll b.1 c.dc cond=neg addr=b
dc = 0000/004e nzcv = 0000

Hit Enter key to step, digit to change step, q to quit: q
```

(Hit Enter key)

(Trace 2 instructions at a time)

The trace shows the before/after contents of registers in the register bank and the after contents of the n, z, c, and v flag registers. For example,

```
ir = 0000/0027 nzcv = 0000
```

shows that the `ir` was changed from 0000 to 0027 and the n, z, c, and v flag registers were all set to 0.

The second debugging aid on `sim` is the `bp` (breakpoint) instruction. When `sim` executes a `bp` instruction, it pauses. If you then hit the Enter key, `sim` continues executing microinstructions. But if you enter a *single* nonzero digit, the step/trace function is activated.

Here is how you might use a `bp` instruction. The `btest.e` program displays the numbers 1 to 10 if the microcode for the basic instruction set is correct. But suppose for your microcode, only the numbers

Chapter 7: Microprogramming the Basic Instruction Set

1 to 4 are displayed. To zero in on the bug in your microcode, put a `bp` instruction immediately after the `dout` instruction that displays 4. Reassemble `btest.a`. Then when `sim` runs `btest.e`, it will pause just before the problem area in your microcode is executed. At the pause, activate the step/trace feature. If you then use the trace to see exactly what is happening in your microcode from that point forward, you should be able to quickly identify the bug.

Compiling C Code to the Basic Instruction Set

To see how good the basic instruction set is, let's see how well is supports code written in C. Let's translate the following C program to the basic instruction set:

```
 1 // e0701.c
 2 #include <stdio.h>
 3 int s;                    // global variable
 4 int sum(int x, int y)
 5 {
 6     return x + y;         // return the sum of of x and y
 7 }
 8 int main()
 9 {
10     s = sum(5, 7);        // call sum passing it 2 and 3
11     printf("%d\n", s);
12     return 0;             // return 0 to startup code
13 }
```

To create an executable file from this program, it has to be compiled to assembly language, assembled to machine language, and linked, all done by a C compiler. The link step combines the machine code produced by the assembly step with *startup code*. Startup code gets control first when the program is invoked on the command line. It calls `main`. When `main` finishes, it returns to startup code. Startup code then returns to the operating system (but our startup code simply halts).

When our C program is executed, startup code gets control first. It calls `main`. `main` calls the `sum` function on line 10, passing it the arguments 5 and 7 to the parameters `x` and `y`. To do this, `main` pushes the arguments in reverse order (first 7, then 5) onto the stack. For example, to push 7 onto the stack, `main` first decrements `sp` to reserve a slot on the stack. Then it stores 7 in that slot using a `str 0` instruction (the most recent slot created on the stack always has the relative address 0):

```
asp -1      ; decrement sp by adding -1 to it
ldi 7       ; load ac with 7
str 0       ; store ac at relative address 0
```

The push of the of the argument 7 creates on the stack the parameter `y`. The subsequent push of the argument 5 creates on the stack the parameter `x`.

Important observation: The calling sequence in the *calling* function creates the parameters in the *called* function by pushing the values of the arguments in the call onto the stack.

On entry into the sum function, the stack looks like this:

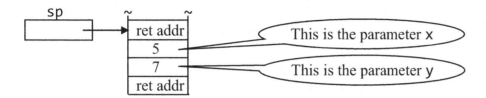

Recall that the stack grows in the downward direction. Thus, the return address pushed by the call instruction in startup code (the first item pushed) is in the location with the highest address, then 7, then 5, then the return address pushed by the call instruction in main. The relative address of y (the slot with 7) is 2; the relative address of x (the slot with 5) is 1. The sum function accesses x and y with two relative instructions:

```
        ldr 1           ; load x
        addr 2          ; add y
```

Note that there are *no x and y labels* in the assembly language program. The parameters x and y are dynamically created (i.e., created during run time) by push operations—not by .fill directives with the labels x and y. After computing the sum of x and y, the sum function returns to main by executing a ret instruction, which pops the return address off the stack and into the pc register. But the x and y parameters are still on the stack. Thus, the first action by main on return from sum is to pop x and y off the stack by adding 2 to the sp register:

```
        asp 2           ; pop x and y parameters
```

main then stores the value returned by sum that is in the ac register into the global variable s:

```
        st s            ; assign the returned value to s
```

Next, main displays the value in s with a dout instruction, and moves the cursor to the next line (the C program does this by calling the C library function printf, but we will simply use a dout, nl sequence):

```
        ld s            ; load ac from s
        dout            ; display value in s
        nl              ; move cursor to next line
```

Here is the assemble code for the entire program, including startup code:

```
1 ; e0701.a
2 startup    call main       ; pushes return address
3           halt
4 ; ================================================
5 s         .fill 0         ; global variable
6 sum       ldr 1           ; load x
7           addr 2          ; add y
8           ret             ; return with sum in ac
9 ; ================================================
```

```
10  main        asp -1          ; decrement sp by adding -1
11              ldi 7           ; load ac with 7
12              str 0           ; store ac at rel addr 0
13              asp -1          ; decrement sp by adding -1
14              ldi 5           ; load ac with 5
15              str 0           ; store ac at rel addr 0
16              call sum        ; pushes return address
17              asp 2           ; pop x and y parameters
18              st s            ; store value returned into s
19              ld s            ; load ac from s
20              dout            ; display value in s
21              nl              ; move cursor to next line
22              ldi 0           ; load ac with 0
23              ret             ; return 0 to startup code
```

Based on this program, the basic instruction set does a good job in supporting C code. However, this program uses only a very small subset of C. It turns out that the basic instruction set has major flaws. It, in fact, does *not* do a good job in supporting C code. In the next chapter, we will microcode a completely different instruction set that is not as flawed as the basic instruction set.

Local variables (i.e., variables declared within a function) in C and C++, like parameters, are dynamically created on the stack. They are *not* created by .fill directives. If a function has local variables, they are created by that function when it is called (by decrementing the sp register), and are destroyed by that function (by incrementing the sp register) just before the function returns to its caller. Let's look at a simple C program that has a local variable y:

```
1  // e0702.c
2  #include <stdio.h>
3  void f(int x)
4  {
5      int y;                  // local variable created here
6      y = x;
7      printf("%d\n", y);
8  }                           // local variable destroyed here
9  int main()
10 {
11     f(3);   // parameter x created here before call, destroyed after call
12     return 0;
13 }
```

The calling sequence in main corresponding to line 11 creates the parameter x by pushing 3 onto the stack. On return from f, the parameter x is destroyed by main by incrementing the sp register. On line 5, the f function creates the local variable y by decrementing the sp register. Just before returning to main, the f function destroys y by incrementing the sp register. Here is the corresponding assembler code:

```
 1 ; e0702.a
 2 startup    call main       ; pushes return address
 3            halt
 4 ; ==============================================
 5 f          asp -1          ; create local variable y
 6            ldr 2           ; get x
 7            str 0           ; store into y
 8            ldr 0           ; get y
 9            dout            ; display y
10            nl
11            asp 1           ; destroy y
12            ret             ; return to main
13 ; ==============================================
14 main       asp -1          ; create x on the stack
15            ldi 3           ; get 3
16            str 0           ; store 3 into x
17            call f
18            asp 1           ; destroy x
19            ldi 0           ; return 0 to startup code
20            ret
```

Local variable y created and destroyed by called function

Parameter x created and destroyed by calling function

Note that there are no labels for the parameter x or the local variable y. Parameters and local variables are not accessed by name (because they have no name at the assembly level) *but by their relative addresses*. For example, on line 6, x in the function f is accessed with a ldr instruction which specifies the relative address 2 (at relative addresses 0 and 1 are the local variable y and the return address from the call instruction, respectively).

Problems

1) Complete the microcode for the basic instruction set in the file b.sm. Test your microcode by entering on the command line

   micro b.sm
   basic btest.a
   sim btest.e

   Prefix these commands with "./" on a Linux, Mac OS X, or Raspbian system. Hand in the ".lst" file created by the micro program and the ".log" file created by the sim program.

2) How does the sim program know it should use the microcode in b.m when it runs the program in btest.e? *Hint*: What is in the first byte of btest.e?

3) Is add the only ALU operation that can be used in the microinstruction at the start of the decoding process that loads the ir from the mdr and sets the flag registers?

4) Is the ld instruction on line 19 in e0701.a necessary?

5) Assemble the following microinstructions to their 40-bit forms. Give your answers in hex.

   ```
   a.pc add b.1 c.pc rd
   a.mdr add b.0 c.ir neg@L1         ; L1 at address 00000000111
   a.ir sll b.1 c.dc neg@L01 ; L01 at address 00000001011
   ```

6) Give the symbolic form of the following microinstructions, which are from the basic instruction set microcode:

   d36d28180b
   000f878000

7) What is wrong with the following microinstruction:

   ```
   a.1 c.mdr rd
   ```

8) Modify the program in e0701.c by making the s variable a local variable within main by moving its declaration to just before the assignment to s in the main function. Local variables are created on the stack by decrementing the sp register. They are accessed using the relative instructions.

9) The symbolic microcode assembler (micro) is case insensitive. The basic assembler is also case insensitive except for labels. Why the difference?

10) Why are the arguments in a function call pushed in right-to-left order? *Hint*: Consider a function with a variable number of parameters (for example, printf in C).

11) Give the assembler code that creates an uninitialized local variable in C? Give the code that creates an initialized local variable in C. Is there a "cost" to initializing local variables?

12) Same as question 11 but for global variables in C.

13) Can the microinstructions that perform the decoding process be restructured so that some of the branching microinstructions can be eliminated, resulting in more efficient microcode?

14) Create a file ce0701.a that contains the basic instruction set assembly code for the following C program. Assemble with the basic program and run with sim. Hand in the ce0701.lst file created by the basic program and the ce0701.log created by sim.

    ```
    // ce0701.c
    #include <stdio.h>
    int y = 7, z;
    int add10(int x)
    {
        return x + 10;
    }
    int main()
    ```

```
   {
      z = add10(y + 3);
      printf("%d\n", z);
      return 0;
   }
```

15) Create a file `ce0702.a` that contains the basic instruction set assembly code for the following C program. Assemble with the `basic` program and run with `sim`. Hand in the `ce0702.lst` file created by the `basic` program and the `ce0702.log` created by `sim`.

```
// ce0702.c
#include <stdio.h>
int a, y = 7;
void f(int x, int y, int z)
{
   int result;
   result = x + y - z;
   printf("%d\n", result);
}
int main()
{
   int b = 7, c;
   a = 1;
   c = a + b + y;
   f(a, b, c);
   return 0;
}
```

16) Create a file `ce0703.a` that contains the basic instruction set assembly code for the following C program. Assemble with the `basic` program and run with `sim`. Hand in the `ce0703.lst` file created by the `basic` program and the `ce0703.log` created by `sim`.

```
// ce0703.c
#include <stdio.h>
void g(int x)
{
   printf("%d\n", x);
}
void f(int x)
{
   g(x - 2);
}
int main()
{
   f(5);
   return 0;
}
```

8 Microprogramming the Stack Instruction Set

Pointers

C and C++ support pointers. A pointer is a variable that contains an address. For example, suppose x is a variable at the memory address 100 that contains the constant 5, and p is a variable that has the address of x:

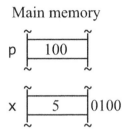

Then we say that p points to x. We typically represent the address in p with an arrow:

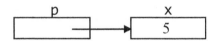

In C and C++, to assign the address of x to p, we use the address-of operator "&":

```
p = &x;    // assign p the address of x
```

To get the address in p, we use p itself. For example, to assign q the address in p, we use

```
q = p;     // assign q the address in p
```

But to access what p is pointing to, we use the dereferencing operator "*". For example, to assign y what p is pointing to, we use

```
y = *p;    // assign y what p is pointing to
```

Thus, if p is pointing to x, and x contains 5, then 5 is assigned to y. If we access what p is pointing to, as in the preceding example, we say we are *dereferencing* p.

We also dereference p to assign a new value to the location p is pointing to. For example, to assign 7 to the memory location p is pointing to, we use

```
*p = 7;    // assign 7 to what p is pointing to
```

Thus, if p is pointing to x, then 7 is assigned to x.

In the examples above, x and y hold integers, and p and q hold pointers to integers. These variables have to be declared appropriately before they are used. Here are the required declarations:

```
int x = 5, y;      // declare x and y to be type int, initialize x to 5
int *p, *q;        // declare p and q to be type int pointer
```

Read "int *" as "integer pointer." Thus, int *p declares p to be an integer pointer. Note that when it appears in a declaration, the asterisk is *not* the dereferencing operator—it is used simply to indicate that a variable is a pointer.

Most programming languages do not support pointers as extensively as C and C++. But virtually all languages use pointers, although their use may be hidden. For example, a reference variable in an object-oriented programming language is a pointer—it points to an object.

Because of the ubiquity of pointers in programming languages, any worthwhile instruction set must provide good support for pointers. Specifically, the machine code for the following C statements should be simple and efficient:

```
p = &x;    // assign p the address of x
*p = 5;    // assign 5 to the location p points to
y = *p;    // assign y what p points to
```

both for variables created with .fill directives and for variables created on the stack (i.e., function parameters and local variables). Let's see how the basic instruction set handles these statements if x, y, and p created with .fill directives:

```
x           .fill 0
y           .fill 0
p           .fill 0
```

It is easy to get the address of x using the basic instruction set. We simply use the ldi instruction:

```
ldi x       ; get address of x
st p        ; store address into p
```

Recall that the assembler translates labels to their corresponding addresses. Thus, the immediate field in the ldi instruction above contains the address of x. When executed, the ldi instruction loads the immediate value it contains (which is the address of x) into the ac register. The st instruction then stores the address into p.

How does the basic instruction set handle dereferencing pointers? We can easily load the address in p into the ac register using a ld instruction (or a ldr instruction if the pointer is on the stack):

```
ld p
```

But we have no way to get to the location at the address in the ac register. Thus, with the basic instruction set, there is *no simple way to dereference pointers*. Moreover, with the basic instruction set, there is no way to get the address of a variable on the stack. For example, suppose x is a parameter or a local variable (in which case it on the stack and is not created with a .fill directive). We have no way of getting its address. For example, suppose x has the relative address of 5. Then its actual 16-bit main memory address is given by the contents of the sp register plus 5. But with the basic instruction set, we have no way of accessing the contents of the sp register. Thus, we have no way of getting the address of a parameter or a local variable. In view of the limitations of the basic instruction set on handling pointers, we conclude that it is an *unacceptable instruction set* for any practical use.

Non-Constant Relative Addresses

Suppose in a C++ program you have the following sequence of statements, where x is a variable on the stack with relative address 2:

```
x = 5;    // relative address of x is 2
int z;    // decrement sp to reserve slot on stack for z
z = x;    // relative address of x is now 3
```

Here is the corresponding assembler code:

```
ldi 5      ; get 5
str 2      ; store 5 in x
asp -1     ; reserve slot on stack for z
ldr 3      ; get x
str 0      ; store in z
```

Wrong relative address?

You might think that there is a bug in the `ldr` instruction above. It is supposed to get x. The relative address of x is 2 but the `ldr` instruction is using the relative address 3. But, in fact, 3 is the correct relative address. The declaration of z causes the `sp` register to be decremented by 1 to reserve a slot on the stack for z. This change in z causes the relative address of each item on the stack to be increased by 1. Thus, the relative address of x changes from 2 to 3.

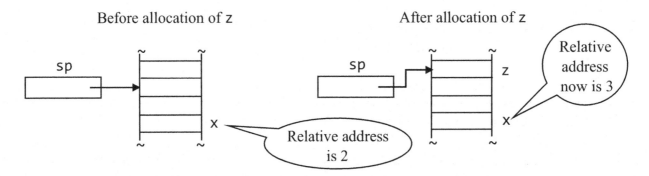

Non-constant relative addresses make writing assembler code or writing compilers that generate assembler code unnecessarily difficult. The solution to this problem is to use two registers—the `sp` register to keep track of the top of the stack and a second register—let's call it `fp` (frame pointer)—to provide the base address for the relative addresses. Then changes to the `sp` register will not affect relative addresses.

Stack Architecture

From the preceding discussion, we can see that the basic instruction set is seriously flawed. Let's now turn our attention to a better instruction set—the stack instruction set—that uses the top of the stack in place of the `ac` register. Unlike the basic instruction set, it can dereference pointers, and relative addresses do not change during the execution of a function.

In the stack instruction set, there is no `ac` register. If, for example, we want to display a value with the `dout` instruction, the value must be on top of the stack. The `dout` instruction then pops and displays that value. To add two values, the two values must be the top two values on the stack. The `add` instruction then pops both values, adds them, and pushes the sum back onto the stack.

Because the stack is central to the operation of the LCC with the stack instruction set, we say that the LCC with the stack instruction set has a *stack architecture*. The LCC with the basic instruction set has a *register architecture* (because the `ac` register is central to its operation). Thus, the same computer can have more than one architecture depending on the instruction set it uses.

Here is the stack instruction set code that pushes 2 and 3, adds them, and displays the result using the `pi` (push immediate), `add`, and `dout` instructions:

```
pi 2    ; push 2 onto the stack
pi 3    ; push 3 onto the stack
add     ; pop twice, add, push sum back onto to stack
dout    ; pop and display sum
```

The `pi` instruction is like the `ldi` instruction in the basic instruction set except that the immediate value in the instruction is pushed onto the stack—not loaded into the `ac` register.

To assign a value to a variable, the top of the stack must have the value and just below it the target address. The `stav` instruction then pops the value and the target address, and then stores the popped value at the popped address. For example, here is the code that assigns 5 to x (where x is a variable created by a `.fill` directive):

```
pi x      ; push address of x onto the stack
pi 5      ; push 5 onto the stack
stav      ; pop twice and store value at address
```

The assembler translates labels to addresses. Thus, in the first `pi` instruction above, the assembler translates the label x to the address of x and places the address in the immediate field of the instruction. Thus, when executed, the `pi` instruction pushes the immediate value (the address of x) onto the stack. Note that the `add` instruction consists of just the mnemonic. The two operands are on the stack and the sum goes back onto the top of the stack. Thus, neither the operands nor the location for the result have to be specified in the instruction.

If x is a variable created on the stack with relative address, say 3, then we assign 5 to it with the following sequence:

```
cora 3    ; convert relative address 3 to absolute address
pi 5      ; push 5 onto the stack
stav      ; pop twice and store value at address
```

The `cora` (convert relative address) instruction specifies a relative address. It converts that relative address to the corresponding absolute address by adding the relative address and the contents of the `fp` register. It then pushes the resulting address onto the stack. Thus, when the `stav` instruction in the sequence above is executed, the top two items on the stack are the value 5 and the absolute address of x. Thus, the `stav` instruction stores 5 in x.

Frame Pointer Register

With the basic instruction set, the sp register has a dual purpose: to keep track of the top of the stack and to provide the base address for the relative instructions. This dual use of the sp register has a major disadvantage: The relative addresses of variables on the stack (parameters and local variables) change whenever a push or pop operation occurs. Non-constant relative addresses make writing assembler code or writing a compiler that generates assembler code unnecessarily difficult. The fix for this problem—which we incorporate in the stack instructions set—is to continue to use the sp register to keep track of the top of the stack but to use a new register—the fp (frame pointer) register—to provide the base address for the relative instructions. Then pushes and pops will not affect relative addresses.

The items on the stack corresponding to the call and execution of a function (the parameters, return address, and local variables) collectively are called a *stack frame*. When a function is executing, the fp register should point to its stack frame.

Consider a program that consists of a main function and an f function. When the main function is executing, the fp register should point main's stack frame. When main calls f, fp should then point to f's stack frame. When f returns to main, the fp register must be restored so that it again points to main's stack frame:

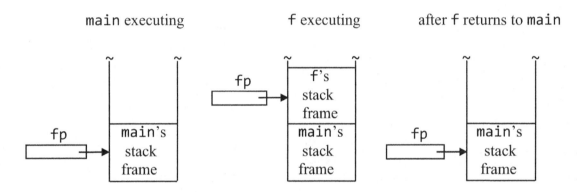

To restore fp so that it points to main's stack frame requires that the fp register be saved when main calls f. The saving of the fp register is performed by the esba (establish base address) instruction. The esba instruction is executed at the very beginning of every function. It saves the frame pointer of its caller by pushing it onto the stack. The restoring of the fp register that occurs when a function is about to return to its caller is performed by the reba instruction. The reba instruction is executed at the end of every function, just before the ret instruction. Here is the form of a program in which main calls f:

```
startup    asp -1 ; reserve space on stack for return code from main
           call main
           halt
; =============
f          esba   ; saves main's frame pointer, loads fp with f's
             ⋮
           reba   ; removes locals (if any), restores main's frame pointer
           ret
; =============
main       esba   ; saves startup's frame pointer, loads fp with main's
             ⋮
           reba   ; removes locals (if any), restores startup's frame pointer
           ret
```

The `reba` instruction also removes local variables, if any, from the stack before a function returns to its caller.

Here is the sequence of events that creates the stack frame for a function:

1. If the called function returns a value, the calling sequence starts by decrementing the `sp` register. This reserves the slot on the stack that receives the value returned by the called function.
2. The calling sequence then pushes the values of the arguments, if any, in right-to-left order, thereby creating the parameters in right-to-left order. Thus, the first parameter is located at the lowest memory address.
3. Next, the calling sequence executes a `call` instruction, which pushes the return address onto the stack and passes control to the called function.
4. The called function executes an `esba` instruction, which saves the frame pointer of the calling function by pushing it onto the stack.
5. Finally, the called function creates its local variables, if any, by decrementing the `sp` register.

Thus, in order of increasing addresses, a stack frame consists of

1. local variables, if any
2. the frame pointer for the calling function
3. return address (i.e., address in the calling function to return to)
4. parameters, if any
5. a reserved slot if the called function returns a value to its caller

During the execution of a function, the `fp` register points to the stack frame for that function—specifically to the slot in its stack frame that holds the frame pointer for the calling function:

102   Chapter 8: Microprogramming the Stack Instruction Set

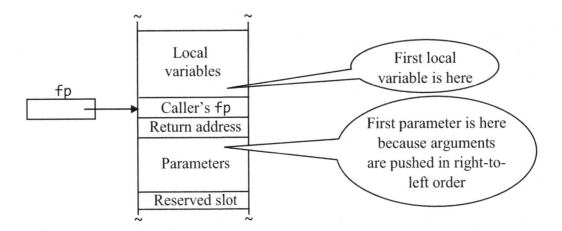

The fp register points *to the middle* of the stack frame, below which are the local variables and above which are the parameters. Thus, local variables have *negative* relative addresses, and parameters have *positive* relative addresses. The relative address of the first local variable is always −1; the relative address of the first parameter is always 2.

Let's examine in detail what happens on the stack when the following program is executed:

```
1  // e0801.c
2  #include <stdio.h>
3  void f(int x)
4  {
5      int y;   // local variable created here
6      y = x;
7      printf("%d\n", y);
8  }   // local variable destroyed here
9  int main()
10 {
11     f(3);   // param x created here before call, destroyed after call
12     return 0;
13 }
```

The program starts with startup code reserving a slot on the stack for the return code that main returns on line 12. It then calls main, which pushes the return address—the address in startup code that main returns to—onto the stack. On entry, main executes an esba instruction which pushes the fp register (which contains 0 because there is no stack frame for our highly simplified startup code) onto the stack. The esba instruction also loads the fp register from the sp register, so that both the sp and fp registers point to the same slot on the stack. The stack at this point looks like this:

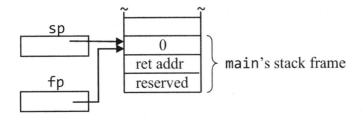

The calling sequence corresponding to line 11 pushes 3 onto the stack (which creates the parameter x). It then calls f with the call instruction, which pushes the return address (the address in main that f returns to) onto the stack. Thus, the stack on entry into f looks like this:

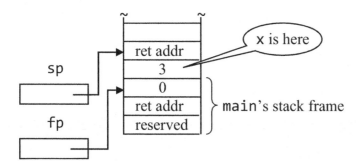

Immediately on entry, f executes an esba instruction. It first saves fp (which contains main's stack frame pointer) by pushing it onto the stack. It then loads fp with the current address in sp. The stack then looks like this:

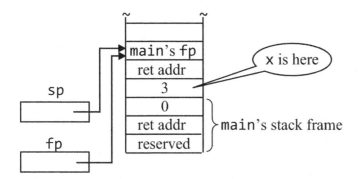

Next, f creates a slot on the stack for the local variable y by decrementing the sp register. We get

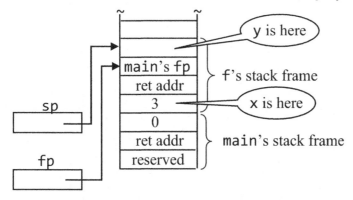

Note that when f is executing, fp does not point to the low end of the stack frame for f. The local variable y is below the location fp points to and the parameter x is above it. Thus, the relative address of y is negative (it is −1), and the relative address of x is positive (it is +2).

Before returning to main, f executes the reba instruction. The reba instruction first loads sp from fp. The effect is to remove the local variable y from the stack. We get

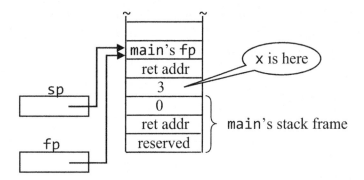

The `reba` instruction then pops the top of the stack into the `fp` register. This has the effect of restoring `fp` with `main`'s stack pointer. We get

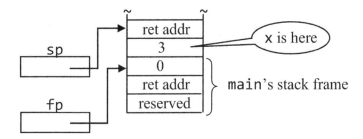

After the `reba` instruction, `f` executes the `ret` instruction. It pops the return address off the stack into the `pc` register which causes a return to `main`. We get

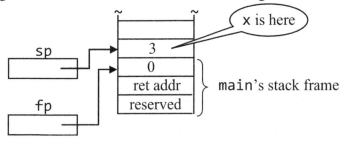

Back in `main`, `main` completes the calling sequence by removing the parameter `x` from the stack by incrementing the `sp` register. We get

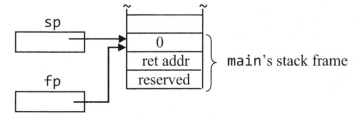

To return to startup code, `main` executes the following sequence of instructions:

```
cora 2 ; push address of reserved slot
pi 0   ; push return code
stav   ; store return code in reserved slot
reba
ret
```

The `cora` instruction pushes the absolute address corresponding to relative address 2 (which is the relative address of the stack location reserved for the return code). The `pi` instruction pushes a 0 return code (0 indicates a normal termination). The `stav` pops the return code and the address of the reserved slot, and then stores the return code at the popped address. The `reba` instruction then restores `fp` with startup code's frame pointer by popping the 0 (which is startup code's `fp` value) into the `fp` register. Finally, the `ret` instruction returns to startup code by popping the return address into the `pc` register. Note that real startup code would then return to the operating system, passing it the return code it received from `main`. But our simplified startup code simply halts.

Here is the assembler code for the program in `e0801.c`:

```
1  ; e0801.a
2  startup    asp -1           ; space for return code
3             call main        ; pushes return address
4             halt
5  ; ============================================
6  f          esba             ; saves main's frame pointer, loads fp with f's
7             asp -1           ; create local variable y
8             cora -1          ; push address of y
9             pr 2             ; push value in x
10            stav             ; assign x to y
11            pr -1            ; push value in y
12            dout             ; display y
13            nl               ; move cursor to next line
14            reba             ; remove y, restore fp
15            ret              ; return to main
16 ; ============================================
17 main       esba             ; saves frame pointer, loads fp with f's
18            pi 3             ; push 3 (creates x)
19            call f           ; push return address, branch to f
20            asp 1            ; destroy x
21            cora 2           ; push address of reserved slot
22            pi 0             ; push 0 return code
23            stav             ; store return code in reserved slot
24            reba             ; restore startup's frame pointer
25            ret              ; return to startup code
```

Dereferencing Pointers

Earlier, we saw that the basic instruction set cannot dereference pointers. Let's see if the stack instruction set does any better. Here is C code that uses pointers and its corresponding assembler code, assuming x, y, and p are created by `.fill` directives:

```
       int x, y, *p;
x          .fill 0
y          .fill 0
p          .fill 0
```

```
p = &x;
    pi p    ; push address of p
    pi x    ; push address of x
    stav    ; pop twice, store address of x in p

*p = 5;
    p p     ; push the pointer in p
    pi 5    ; push 5
    stav    ; pop twice, store 5 where p points
```

In the following sequence, the **dp** (dereference pointer) instruction replaces the pointer on top of the stack with what the top of the stack is pointing to:

```
y = *p;
    pi y    ; push the address of y
    p p     ; push pointer in p
    dp      ; dereference the pointer on top of the stack
    stav    ; pop twice, store what p points to in y
```

All the assembler code sequences to handle pointers are simple and efficient. Now let see if we get the same result if x, y, and p are variables on the stack. Suppose x, y, and p are local variable with relative addresses −1, −2, and −3, respectively.

```
int x, y, *p;
    asp -1      ; create x, relative address = -1
    asp -1      ; create y, relative address = -2
    asp -1      ; create p, relative address = -3

p = &x;
    cora -3     ; push address of p
    cora -1     ; push address of x
    stav        ; pop twice, store address of x in p

*p = 5;
    pr -3       ; push the pointer in p
    pi 5        ; push 5
    stav        ; pop twice, store 5 where p points

y = *p;
    cora -2     ; push the address of y
    pr -3       ; push pointer in p
    dp          ; dereference the pointer on top of the stack
    stav        ; pop twice, store what p points to in y
```

Here again, all the assembler code sequences to handle pointers are simple and efficient. Thus, the stack instruction set fixes two of the major flaws in the basic instruction set:

1. the inability to handle pointers
2. non-constant relative addresses

Multiplying

The stack instruction set has two multiply instructions that work in exactly same way but are implemented differently. mhw (multiply by hardware) uses the multiply circuit within the ALU. mmc (multiply by microcode) uses a procedure in microcode consisting of shift and add operations. A real computer, of course, would not have two instructions that provide exactly the same function. But the LCC does to illustrate an important point: A computational capability can be implemented in hardware. But it can also be implemented in microcode. The hardware approach is the faster but more expensive.

Like the add instruction, the multiply instructions operate on the two numbers on top of the stack. To multiply 2 and 3 with the mhw instruction, we push 2 and 3 with pi instructions, then execute the mhw instruction. It pops the 2 and the 3, multiplies them, and pushes the result back onto the stack. We can then pop and display the result with the dout instruction:

```
pi 2    ; push 2
pi 3    ; push 3
mhw     ; pop twice, multiply, push result
dout    ; pop result and display
nl      ; move cursor to start of next line
```

Here is the microcode for the add instruction in the stack instruction set:

```
a.sp c.mar              ; get address of top of stack into mar
a.sp add b.1 c.sp rd    ; read top of stack and increment sp
a.mdr c.temp            ; save in temp
a.sp c.mar              ; prepare to read 2nd operand, don't incr sp
rd                      ; read 2nd operand into mdr
a.mdr add b.temp c.mdr  ; add, put sum in mdr
wr br@fetch             ; replace 2nd operand with sum in mdr
```

It pops the first operand of the add operation off the stack. The second operand is then on the top of the stack. Because it will be replaced by the sum, it is unnecessary (and inefficient) to pop it off of the stack. Thus, the sp register is not incremented a second time. The microcode for the mhw instruction is the same except that the add ALU operation is replaced with the mult ALU operation.

The microcode for the mmc instruction is more complex. It uses a loop that repeatedly performs add and shift operations. It multiplies the same way we multiply with a pencil and paper. For example, let's multiply 00011 (3 decimal) by 00101 (5 decimal). We get the product 01111 (15 decimal):

```
          00011 multiplicand
          00101 multiplier
          ─────
          00011  ⎫
         00000   ⎪
        00011    ⎬ Partial products
       00000     ⎪
      00000      ⎭
      ─────────
      000001111 product
```

The multiplicand (the top number) is multiplied by each bit of the multiplier (the bottom number). The partial products are either equal to the multiplicand (if the multiplying bit is 1) or 0 (if the multiplying bit is 0). The sum of the partial products appropriately shifted is the product. We can perform this process in microcode with a loop that adds and shifts. Here is the pseudocode for the microcode we need:

```
1                 pop stack into temp (temp is the multiplier)
2                 pop stack into mdr (mdr is the multiplicand)
3                 set ac to 0
4  loop           if temp (the multiplier) = 0 branch to done
5                 if rightmost bit of temp (the multiplier) = 0 branch to skip
6                 add mdr (the multiplicand) to ac
7  skip           shift temp (the multiplier) right one position
8                 shift mdr (the multiplicand) left one position
9                 branch to loop
10 done           push ac (the product)
11                branch to fetch
```

Each time through the loop, the `temp` register (the multiplier) is shifted right one position (line 7). Thus, the test of the rightmost bit of `temp` on line 5 tests a different bit in the multiplier each time it is executed. The `mdr` (the multiplicand) is shifted left on line 8 each time through the loop. Thus, whenever it is added to the `ac` register on line 6, it is in the appropriately shifted position. At the conclusion of the loop, the `ac` register holds the product. The `ac` register here is used simply as another temporary register in which to accumulate the sum of the multiplicands. It is not accessible at the machine level. In place of the `ac` register, we could have used any other available register, such as `r10`.

Adding Opcodes

With four bits to represent opcodes, we can represent only 16 opcodes. However, some of the instructions in the stack instruction set do not specify any operands. Thus, their rightmost 12 bits are unused and, therefore can be used as an extension of the opcode. For example, in the stack instruction set, dp, esba, reba, mhw, and mmc all have the four-bit opcode 0101 (5 hex), but have different opcode extensions in their rightmost 12 bits:

	Opcode (hex)	Opcode Extension (hex)
dp	5	000
esba	5	001
reba	5	002
mhw	5	004
mmc	5	008

The opcode extensions for these instructions are decoded in microcode with a succession of shift right operations that test the bits in the opcode extension from right to left. Each right shift operation shifts the register specified by the A field. The rightmost bit of the register is shifted into the carry flag. The conditional branch in each shift instruction branches if the bit shifted into the carry flag is 1:

```
L0101      ; dp, esba, reba, mhw, mmc ===========================
           a.ir and b.m12 zer@dp     ; branch if opcode extension = 0
           a.ir srl b.1 c.dc cy@esba ; branch if bit 0 in opcode ext = 1
           a.dc srl b.1 c.dc cy@reba ; branch if bit 1 in opcode ext = 1
           a.dc srl b.1 c.dc cy@mhw  ; branch if bit 2 in opcode ext = 1
           a.dc srl b.1 c.dc cy@mmc  ; branch if bit 3 in opcode ext = 1
           br@fetch
dp         ; microcode that interprets dp
              ⋮
esba       ; microcode that interprets esba
              ⋮
reba       ; microcode that interprets reba
              ⋮
mhw        ; microcode that interprets mhw
              ⋮
mmc        ; microcode that interprets mmc
              ⋮
```

Stack Instruction Set Summary

Opcode	Format		Description
0	p	x	mem[--sp] = mem[x];
1	pi	x	mem[--sp] = x;
2	pr	s	mdr = mem[fp + s]; mem[--sp] = mdr;
3	cora	s	temp = fp + s; mem[--sp] = temp;
4	stav		temp = mem[sp++]; mem[mem[sp++]] = temp;
5000	dp		mem[sp] = mem[mem[sp]];
5001	esba		mem[--sp] = fp; fp = sp;
5002	reba		sp = fp; fp = mem[sp++];
5004	mhw		temp = mem[sp++]; mem[sp] = mem[sp] * temp;
5008	mmc		temp = mem[sp++]; mem[sp] = mem[sp] * temp;
6	asp	s	sp = sp + s;
7	add		temp = mem[sp++]; mem[sp] = mem[sp] + temp;;
8	sub		temp = mem[sp++]; mem[sp] = mem[sp] - temp];
9	call	x	mem[--sp] = pc; pc = x;
A	ret		pc = mem[sp++];
B	br	x	pc = x;
C	brn	x	if (mem[sp++] < 0) pc = x;
D	brz	x	if (mem[sp++] == 0) pc = x;
E	brp	x	if (mem[sp++] > 0) pc = x;
F	trap	y	see below

		Description
halt	or trap 0	Terminate program
nl	or trap 1	Output nl character
dout	or trap 2	Pop and display signed number in decimal
udout	or trap 3	Pop and display unsigned number in decimal
hout	or trap 4	Pop and display number in hex
aout	or trap 5	Pop and display ASCII character
sout	or trap 6	Pop address and display string at that address
din	or trap 7	Read decimal number from keyboard and push
hin	or trap 8	Read hex number from keyboard and push
ain	or trap 9	Read character from keyboard and push
sin	or trap 10	Pop address and read in string to that address
bp	or trap 14	Breakpoint

x: bits 0 to 11 in machine instruction zero-extended to 16 bits
s: bits 0 to 11 in machine instruction sign-extended to 16 bits
y: bits 0 to 7 in machine instruction zero-extended to 16 bits
sp: stack pointer
fp: frame pointer register
mdr: memory data register
temp: temporary register

Directives: .blkw, .fill, .start, .stringz

In our preceding discussion, we saw that the stack instruction set can handle pointers, but the basic instruction set cannot. Moreover, the stack instruction set has the `brp` (branch on positive) instruction as well of the `brn` and `brz` instructions. The basic instruction set has only the `brn` and `brz` branching instructions. Thus, the stack instruction set provides more function than the basic instruction set even though its instructions have the same number of bits (16 bits). How is this possible? By comparing the two instruction sets, we find the answer: To perform an add operation or to perform a store operation, the basic instruction set needs two instructions: `add` or `addr` to add and `st` or `str` to store. But the stack instruction set needs only one: `add` to add and `stav` to store. Moreover, the stack instruction set has more instructions amenable to opcode extensions—that is, more instructions which do not use their rightmost 12 bits, which therefore can be used for opcode extensions.

There is, however, a downside to the stack instruction set. Its instructions use the stack which is in main memory. Accessing memory takes considerably more time that accessing registers. Let's compare the code sequences for the two instruction sets that add `x` and `y` and store the result in `z`:

Basic Instruction Set		Stack Instruction Set	
	Memory accesses		Memory accesses
ld x ;	1	pc z ;	1
add y ;	1	p x ;	2
st z ;	1	p y ;	2
		add ;	3
		stav ;	3

The basic instruction set code makes three main memory accesses plus three more to fetch the instructions. The stack instruction set code makes 11 memory accesses plus five more to fetch the instructions. Six memory accesses total for the basic instruction set versus 16 total for the stack instruction set. Quite a substantial difference! The inherent run-time inefficiency of a stack architecture is the reason why most computers have a register architecture rather than a stack architecture.

Problems

1) Complete the microcode for the stack instruction set in the file named `s.sm`. Test your microcode by assembling the program in `stest.a` by entering

    stack stest.a

    Assemble your microcode by entering

    micro s.sm

    Then run the `stest` program using the simulator program by entering

    sim stest.e

    The `stest` program should display the numbers 1 to 10. Prefix the commands above with "./" on a Linux, Mac OS X, or Raspbian system. Hand in the ".lst" file created by the `micro` program and the ".log" file created by the `sim` program.

2) Do the multiply instructions work for both positive and negative numbers? Confirm you answer by running several test cases.

3) When two 16-bit signed numbers are multiplied, what is the maximum number of bits required by the product?

4) The opcode extensions for the machine instructions with opcode 5 are 000, 001, 002, 004, 008. Why not 000, 001, 002, 003, 004?

5) What stack assembler code corresponds to the following C statement, where p is a variable created with a .fill directive:

y = **p;

Hint: **p accesses the location pointed to by the location p points to. p is a *pointer to a pointer*. It should be declared this way:

int **p;

This declaration consists of two parts: "int *"and "*p". Read "*p" as "p is a pointer" and "int *" as "to an int pointer." Thus, p is a pointer to an int pointer.

6) What stack assembler code corresponds to the following C statement, where p is a variable created with a .fill directive:

y = **p + 1;

7) What assembler code corresponds to the following C statement, where p is a variable created with a .fill directive:

y = *(*p + 1);

8) What stack assembler code corresponds to the following C statement, where p is a variable created with a .fill directive:

y = **(p + 1);

9) What stack assembler code corresponds to the following C statement, where p is a variable created with a .fill directive:

**p = 5;

10) What stack assembler code corresponds to the following C statement, where p is a variable created with a .fill directive:

*(*p + 1) = 5;

11) What stack assembler code corresponds to the following C statement, where p is a variable created with a .fill directive:

    **(p + 1) = 5;

12) Create a file ce0801.a that contains the stack instruction set assembly code for the following C program. Assemble with the stack program and run with sim. Hand in the ce0801.lst file created by the stack program and the ce0801.log created by sim.

    ```
    // ce0801.c
    #include <stdio.h>
    int y = 7, z;
    int add10(int x)
    {
        return x + 10;
    }
    int main()
    {
        z = add10(y + 3);
        printf("%d\n", z);
        return 0;
    }
    ```

13) Create a file ce0802.a that contains the stack instruction set assembly code for the following C program. Assemble with the stack program and run with sim. Hand in the ce0802.lst file created by the stack program and the ce0802.log created by sim.

    ```
    // ce0802.c
    #include <stdio.h>
    int a, y = 7;
    void f(int x, int y, int z)
    {
        int result;
        result = x + y - z;
        printf("%d\n", result);
    }
    int main()
    {
        int b = 7, c;
        a = 1;
        c = a + b + y;
        f(a, b, c);
        return 0;
    }
    ```

14) Create a file ce0803.a that contains the stack instruction set assembly code for the following C program. Assemble with the stack program and run with sim. Hand in the ce0803.lst file created by the stack program and the ce0803.log created by sim.

```
// ce0803.c
#include <stdio.h>
void g(int x)
{
    printf("%d\n", x);
}
void f(int x)
{
    g(x - 2);
}
int main()
{
    f(5);
    return 0;
}
```

15) Create a file `ce0804.a` that contains the stack instruction set assembly code for the following C program. Assemble with the `stack` program and run with `sim`. Hand in the `ce0804.lst` file created by the `stack` program and the `ce0804.log` created by `sim`.

```
// ce0804.c
#include <stdio.h>
int y = 7;
void f(int *p)
{
    *p = *p + 1;
}
int main()
{
    f(&y);
    printf("%d\n", y);
    return 0;
}
```

9 Microprogramming the Optimal Instruction Set

Flaws in the Stack Instruction Set

Although the stack instruction set is an improvement over the basic instruction set, it still has some flaws. For example, the address fields in the stack instruction set, like the address fields in the basic instruction set, are 12 bits wide. With 12 bits, only the lowest 4K (4096) of memory can be addressed. The stack grows downward from the top of main memory. But the size of the stack at its largest is generally very small—typically less than 20 words. Thus, a program written with the stack instruction set (or the basic instruction set) can take advantage of only a small portion of the memory available on the LCC.

A more serious flaw in both instruction sets is their inability to correctly compare two signed numbers. To compare two numbers, the bottom number is subtracted from the top number. If the result is negative, then the top number is less than the bottom number. For example, to determine if x is less than y with the stack instruction set, we push x, push y, and then subtract. If the result is negative, then x < y. In the following code, a branch to less occurs if x < y:

```
p x         ; push x
p y         ; push y
sub         ; compute x - y
brn less    ; branch if x < y
```

Unfortunately, this approach does not work if overflow occurs during the subtraction. For example, suppose x is 0111111111111111 (32767 decimal) and y is 1111111111111111 (−1 decimal). To subtract, the LCC adds the two's complement of y to x:

```
  0111111111111111  (32767)
+ 0000000000000001  (two's complement of −1)
  1000000000000000  (−32768)
```

The result is negative, which *incorrectly* implies x (which is 32767) is less than y (which is −1). The result of the subtraction should be +32768 (note that 32767 − (−1) = +32768). But because this value is out of the range of 16-bit two's complement numbers, the computed result is incorrect, which results in an incorrect compare operation. Moreover, we do not have any easy way with both the stack and basic instruction sets to determine if overflows occurs. Thus, we cannot easily modify our approach to comparing numbers so that it always yields a correct compare.

Out next and last instruction set, the optimal instruction set, fixes both of the flaws discussed above. We call it the optimal instruction set because it is genuinely optimal for the purpose for which it was designed: namely, to see what C and C++ code looks like at the assembly level. It is the only instruction set used in the companion volume for this book, *C and C++ Under the Hood*. It is the standard instruction set for the LCC, and therefore is generally called the "LCC instruction set." However, in this book, we call it the "optimal instruction set" to distinguish it from the basic and stack instruction sets.

In this chapter, we focus on the microprogramming of the optimal instruction set —not on using it to write programs. For extensive coverage on using the optimal instruction set, see *C and C++ Under the Hood*.

Instruction Set Architecture

The *instruction set architecture* (ISA) of a computer is the collection of the features of the computer that are "visible" at the machine-instruction level. For example, the ISA of the LCC with the basic instruction set consists of the 64K main memory, the ac, pc, and sp registers, and its machine instruction set. Let's now look at the ISA of the LCC with the optimal instruction set. It consists of the 64K main memory, eight registers, named r0 to r7, the pc and flag registers, and its machine instruction set. Registers r5, r6, and r7 have the alternate names fp (frame pointer), sp (stack pointer), and lr (link register), respectively:

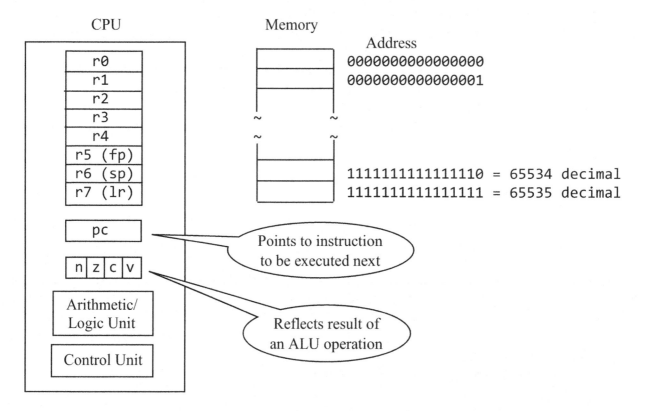

The optimal instruction set uses *pc-relative addresses* (also called *pc-offsets*). That is, the addresses within instructions are relative to the location that the pc register is pointing to. When executed, an instruction with a pc-relative address accesses the memory location whose address is given by the current address in the pc register plus the pc-relative address in the instruction:

(address in pc) + (pc-relative address)

We call result resulting address the *absolute* or *effective address*. It is the address that is actually sent to the memory unit on a read or write operation. For example, if an instruction at address 2 has a pc-relative address 5, the absolute address is 3 + 5 = 8. That is, the instruction accesses the memory location at the address 3 + 5 = 8. Why 3 + 5 and not 2 + 5? Recall that the CPU increments the pc after it fetches an instruction and before it executes the instruction. When an instruction with a pc-relative address is executed, the pc register has already been incremented so that it is pointing to the next instruction. Thus, a pc-relative address in an instruction is not relative to the address of the instruction but to the next address.

The assembler for the optimal instruction set computes `pc`-relative addresses for us—we do not have to do it. For example, consider the following code:

```
        ld r1, x
          ⋮
x       .fill 5
```

Suppose the `ld` (load) instruction is at location 2 and x is at location 8. The assembler determines the `pc`-relative address by computing

$$\text{pc-relative address} = (\text{address of x}) - (\text{address of ld} + 1) = 8 - (2 + 1) = 5$$

A relative address is the number of words between two points in a program (from the location right after the instruction to the target label). It *does not depend on the load point*. Thus, unlike the absolute addresses in the basic and stack instruction sets, relative addresses do not have to be adjusted according to the load point. If a program with the code above were loaded into memory so that the `ld` instruction is at address 1002, then x would be at the address 1008. The correct relative address is still 5:

1003 (address in `pc` when the `ld` executed) + 5 (relative address) = 1008 (address of x)

The `ld` instruction is one of the instructions in the optimal instruction set that has a `pc`-relative address. Here is its format:

    0010 C pcoffset9

It consists of a four-bit opcode (0010), the C field (the three-bit number of the destination register), and pcoffset9 (a nine-bit `pc`-relative address). For example, if the `pc`-relative address of x in the following instruction is 5,

        ld r1, x

then its machine language version is

    opcode C   pcoffset9
    0010   001 000000101

or 2205 in hex. The pcoffset9 field is treated as a signed number. Thus, it can be positive, zero, or negative. It can range from −256 to 255. The `st` (store) instruction has the same format as the `ld` instruction.

The `lea` (load effective address) instruction has the same format as the `ld` instruction. When executed, it loads the destination register with the absolute address corresponding to its pcoffset9 relative address. That is, it loads the destination register with

    (contents of pc register) + pcoffset9

Thus, the `ld` instruction

        ld r0, x

loads r0 with the value at x, but

>   lea r0, x

loads r0 with the 16-bit main memory address of x.

The ldr (load relative) instruction loads the destination register from the address given by a register plus an offset (which can be positive, zero, or negative. For example,

>   ldr r3, fp, -2

load r3 from the address given by the contents of fp plus −2. Here is its format:

>   0110 C A offset6

Its C field contains the number of the destination register; its A field contains the number of the register with the address. For example, the assembler instruction above is translated to

>   0110 011 101 111110

The offset6 field is six bits wide. Its range is from −32 to +31. The str (store relative) instruction has the same format at the ldr instruction.

The mvi (move immediate) instruction moves its nine-bit immediate value to the destination register. Here is its format:

>   1101 C imm9

For example,

>   mvi r3,-2

moves −2 into r3. The imm9 can range from −256 to +255

The add instruction has two formats:

>   0001 C A 000 B

and

>   0001 C A 1 imm5

The C, A, and B fields are all three-bit register fields. The B field is in positions 2 to 0 in the instruction, the A field in positions 8 to 6, and the C field is in positions 11 to 9. Here is an assembler instruction that is translated to the first format:

>   add r0, r1, r2

It adds the contents of r1 and r2 (the registers specified by the A and B fields) and loads r0 (the register specified by the C field) with the sum. The second format includes an immediate value in place of the third register:

>   add r0, r1, -3

This instruction adds the contents of r1 and the immediate value −3, and loads r0 with the sum. The immediate value is in the imm5 field of the instruction. Because imm5 is only 5 bits wide, its range is from only −16 to +15.

Both the sub (subtract) and the and instructions have the two formats that are the same as the two formats for the add instruction. The and instruction performs a bitwise AND. The cmp instruction subtracts just like the sub instruction. But unlike the sub instruction, it does not load a destination register with the result (its only function is to set the flag registers). Thus, its formats are the same as the formats for the sub instruction except that the C field is not used (it contains 000)

The jsr (jump to subroutine) jumps (i.e., branches) to the specified subroutine after storing the return address in the link register lr (r7). Here are its two formats:

   0100 1 pcoffset11

and

   0100 000 A 000000

In the first format, the jump-to address is provided by the 11-bit pc-relative address in the instruction. In the second format, the jump-to address is provided by the register specified by the A field in the instruction. Here is an assembler instruction that is translated to the first format:

```
        jsr sub     ; load lr with return address, jump to sub
```

An assembler instruction that is translated to the second format specifies a register:

```
        jsr r1      ; load lr with return address, jump to address in r1
```

Because the jsr instruction loads lr with the return address (it does not push it onto the stack like the call instruction in the basic and stack instruction sets), the ret instruction returns simply by copying the contents of lr to the pc register.

The not instruction performs a bitwise NOT operation. It makes a copy of the contents of the register specified by its A field, performs a bitwise NOT operation on it, and moves the result into the destination register specified by its C field. For example,

```
        not r1, r2
```

moves the bitwise NOT of the contents of r2 into r1.

The push instruction pushes the contents of the specified register onto the stack. The pop instruction pops the top of the stack into the specified register. For example, the sequence

```
        push r1     ; push contents of r1 onto stack
        pop r2      ; pop top of the stack into r2
```

pushes the contents of r1 onto the stack and then pops the top of the stack into r2.

The br (branch) instructions branch to the specified address if the condition specified by the code in its C field is true. Here is its format:

   0000 code pcoffset9

The branching condition is specified at the assembly level by extending the br mnemonic with a suffix that represents the branching condition. For example

        `brn loop`   ; branch on negative

branches on negative to the label loop. The code for brn is 010. Thus, its machine instruction would have 010 its code field. Here is a summary of the branching mnemonics, along with their codes and the condition tested:

Mnemonic	Code	Condition Tested
brz or bre	000	z = 1
brnz or brne	001	z = 0
brn	010	n = 1
brp	011	n = z
brlt	100	n != v
brgt	101	n = v and z = 0
brc	110	c = 1
br	111	always branch

The bre, brne, brlt, and brgt instructions are typically used right after a cmp (or sub) instruction. The cmp instruction (the two-register form) subtracts the contents of the register specified by its B field from the register specified by its A field, and sets the flag registers n, z, c, and v to reflect the result. The branching instruction then branches or not depending on the flag registers. For example, here is a sequence that branches to dog if the contents of r1 is less than r2:

```
cmp r1, r2   ; subtract r2 from r1, set flags
brlt dog     ; branch if r1 < r2
```

The branch to dog occurs if the condition for the brlt instruction, n != v, is true. If n != v, then either n = 1 and v = 0 or n = 0 and v = 1.

    Case 1: n = 1 and v = 0
    v = 0 so signed overflow did not occur when the cmp instruction subtracted r2 from r1. Thus, n = 1 indicates that the true result of the subtraction is negative. A negative result in a subtraction indicates that the top number (the contents of r1 in this example) is less than the bottom number (the contents of r2).

    Case 2: n = 0 and v = 1
    v = 1 so signed overflow occurred. Thus, n reflects the sign of the computed result—not the true result. The sign of the true result is 1, which indicates that the true result of the subtraction is negative, again indicating that the top number is less than the bottom number.

Thus, the brlt instruction works correctly whether or not overflow occurs when the cmp or sub instruction that precedes it is executed. Similarly, brgt works correctly whether or not overflow occurs in the cmp or sub instruction that precedes it.

    The basic and stack instruction sets also have no easy way to compare unsigned numbers. But we can do so easily with the optimal instruction set. Recall from chapter 1 that the c flag acts like a borrow flag on a subtraction. In a subtraction of two unsigned numbers, if the top number less than the bottom number, then a borrow into the leftmost position occurs, resulting is the setting of the c flag register to 1. Thus, if c = 1 after a cmp or sub instruction, the top number is less than the bottom number. For example, in the following sequence, suppose r0 and r1 hold unsigned numbers. Then the branch to bird occurs if the r0 number is less than the r1 number:

```
cmp r0, r1  ; unsigned compare
brc bird    ; branch to bird if r0 < r1
```

The optimal instruction set has three types of instructions that transfer control: the branching instructions, the jsr instruction, and the jmp instruction. The jmp instruction unconditionally jumps to the absolute address in the specified register. For example, the following instruction jumps to the absolute address in r3:

```
jmp r3
```

Recall that the two principal flaws in the stack instruction set (which are also in the basic instruction set) are its inability to correctly compare numbers and its inability to access the full 64K of main memory. Our optimal instruction set, however, has neither of these flaws. With a cmp instruction followed by a conditional branching instruction, we can easily compare numbers, both signed and unsigned. Because the addresses within instructions are pc-relative addresses, an optimal instruction set program can be loaded anywhere into memory and still work correctly. If a program is loaded into high main memory, it can still directly access its data and call its subroutines, assuming they are in range of its pc-relative addresses. Even if data or subroutines are out of range, they can still be accessed. For example, suppose the data at x is out of range. We can access it with

```
        ld r1,@x         ; load r1 with address of x
        ldr r0, r1, 0    ; load r0 from x
          :
@x      .fill x          ; @x sub within range of the ld instruction
```

At run time, the location corresponding to the label @x will have the 16-bit absolute address of x. The ld instruction loads this address int r1. Then the ldr instruction loads from the address in r1 to get x. As long as @x is within range of the pc-relative address in the ld instruction, this sequence will work. The label x *can be anywhere*. We can similarly call a subroutine regardless of its location with the following sequence:

```
        ld r1, @sub      ; load r1 with address of sub
        jsr r1           ; jump to subroutine whose address in r1
          :
@sub    .fill sub        ; @sub is within range of the ld instruction
```

We can also unconditionally jump to any location with

```
        ld r1, @abort    ; load r1 with address of sub
        jmp r1           ; jump to address in r1
          :
@abort  .fill abort      ; @abort is within range of the ld instruction
```

Here is a summary of the optimal instruction set:

Mnemonic	Format				Flags	Description
br--	0000	code	pcoffset9			on cond, pc = pc + pcoffset9
add	0001	C	A 000 B		nzcv	C = A + B
add	0001	C	A 1 imm5		nzcv	C = A + imm5
mov	0001	C	A 1 00000		nzcv	C = A
ld	0010	C	pcoffset9			C = mem[pc + pcoffset9)
st	0011	C	pcoffset9			mem[pc + pcoffset9] = C
jsr	0100	1	pcoffset11			lr= pc; pc = [pc + pcoffset11]
jsr	0100	000	A 000000			lr = pc; pc = A
and	0101	C	A 000 B		nz	C = A & B
and	0101	C	A 1 imm5		nz	C = A & imm5
ldr	0110	C	A offset6			C = mem[A + offset6]
str	0111	C	A offset6			mem[A + offset6] = C
cmp	1000	000	A 000 B		nzcv	A - B (set flags)
cmp	1000	000	A 1 imm5		nzcv	A - imm5 (set flags)
not	1001	C	A 000000		nz	C = ~A
push	1010	C	000000001			mem[--sp] = C
pop	1010	C	000000010			C = mem[sp++];
srl	1010	C	000000100		nzc	C >> 1 (0 inserted)
sra	1010	C	000001000		nzc	C >> 1 (sign replicated)
sll	1010	C	000010000		nzc	C << 1 (0 inserted)
sub	1011	C	A 000 B		nzcv	C = A - B
sub	1011	C	A 1 imm5		nzcv	C = A - imm5
jmp	1100	000	A 000000			pc = A
ret	1100	000	111 000000			pc = lr
mvi	1101	C	imm9			C = imm9
mov	1101	C	imm9			C = imm9
lea	1110	C	pcoffset9			C = pc + pcoffset9
trap	1111	0000	trapvect8			see below

halt	or trap 0	Terminate execution		br mnemonic	code
nl	or trap 1	Output newline		brz/bre	000
dout	or trap 2	Display signed number in r0 in dec		brnz/brne	001
udout	or trap 3	Display unsigned number in r0 in dec		brn	010
hout	or trap 4	Display number in r0 in hex		brp	011
aout	or trap 5	Display ASCII character in r0		brlt	100
sout	or trap 6	Display string r0 points to		brgt	101
din	or trap 7	Read dec number from keybd into r0		brc	110
hin	or trap 8	Read hex number from keybd into r0		br	111
ain	or trap 9	Read character from keybd into r0			
sin	or trap 10	Input string into buf r0 points to			
bp	or trap 14	Breakpoint			

Directives: .blkw, .fill, .start, .stringz

Note that the `mov` instruction right below the `add` instruction in the summary above is really just an `add` instruction in which the immediate value is 0. For example, the `mov` instruction

```
mov r1, r2          ; move contents of r2 into r1
```

is really this `add` instruction

```
add r1, r2, 0
```

which adds the contents of `r2` and 0 and moves the result in to `r1`. Thus, its effect is to simply move the contents of `r2` into `r1`. The `mov` instruction is not distinct instruction but just an easy way to represent another instruction. For this reason, we call it a *pseudo-instruction*.

If the second operand in a `mov` instruction is a register, it is translated to an `add` instruction, as illustrated above. However, if the second operand is an immediate value, then it is translated to a `mvi` instruction. For example,

```
mov r1, 3
```

is really the instruction

```
mvi r1, 3
```

Thus, whenever we have to do a move operation, we can simple use the "`mov`" mnemonic followed by the appropriate operands—either two registers or a register and an immediate value.

Programming with the Optimal Instruction Set

To get a feel for the optimal instruction set, let's examine an assembler program that passes and dereferences addresses. Before you study this program, be sure you understand how the `ldr` instruction works. It and the `str` instruction are the instructions that dereference pointers. For example, suppose `r0` contains a pointer. Then

```
ldr r0, r0, 0
```

loads `r0` from the address given by its second are third operands (`r0` and 0). The address given by the contents of `r0` plus 0 is, of course, just the address in `r0`. Thus, the instruction loads `r0` from the location `r0` is pointing to. In other words, it dereferences the pointer in `r0`. The effect of this instruction is to replace the pointer in `r0` with the value the pointer is pointing to.

In the following program, `main` passes the address of `x` to the parameter `p` in the function `f`. `f` then dereferences `p` twice—once to access the location `p` points to (to get the value in `x`) and once to store a new value in the location `p` points to (to store a new value in `x`). The effect is to increment `x` by 1. The comments in the program show the corresponding C code.

124   Chapter 9: Microprogramming the Optimal Instruction Set

```
1  ; e0901.a
2  startup     jsr main
3              halt
4  ; ===============================================
5  x           .fill 7                 ; int x = 7;
6  f           push lr                 ; void f(int *p)
7              push fp,                : {
8              mov fp, sp
9
10             mov r0, 1               ;    int y = 1;
11             push r0
12
13             ldr r0, fp, 2           ;    *p = *p + y;
14             ldr r0, r0, 0
15             ldr r1, fp, -1
16             add r0, r0, r1
17             ldr r1, fp, 2
18             str r0, r1, 0
19
20             mov sp, fp              ; }
21             pop fp
22             pop lr
23             ret
24 ; ===============================================
25 main        push lr                 ; int main()
26             push fp                 ; {
27             mov fp, sp
28
29             lea r0, x               ;    f(&x);
30             push r0
31             jsr f
32             add sp, sp, 1
33
34             ld r0, x                ;    printf("%d\n", x);
35             dout
36             nl
37
38             mov r0, 0               ;    return 0;
39             mov sp, fp
40             pop fp
41             pop lr
42             ret
```

Callouts: "Load from p" (line 13), "Dereferences pointer in r0" (line 14), "Load from p" (line 17), "Dereferences pointer in r1" (line 18), "Creates the parameter p" (lines 29–30), "Removes p from the stack" (line 32).

On line 29, the `lea` instruction loads the address of x. The `push` on line 30 then pushes it, thereby creating the parameter p. Thus, the parameter p points to x. On line 31, the `jsr` loads the return address in the `pc` register into `lr` and then jumps to f. On line 6, the `push` instruction saves the return address in `lr` by pushing it onto the stack. Then, if the address in `lr` is corrupted during the execution of f (it is not for this function), the return address is still available on the stack.

The two instructions on line 7 and 8 together do what the `esba` instruction does in the stack instruction set: They save the caller's frame pointer and then load `fp` with the pointer to the stack frame of the called function (by loading `fp` from `sp`).

Line 11 creates and initializes the local variable `y` by pushing 1 onto the stack. The stack at this point looks like this:

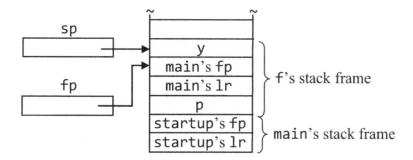

The parameter `p` has the relative address 2 (relative to the address in `fp`). The local variable `y` has the relative address −1. Thus, the `ldr` instruction on line 13 loads `r0` from `p` (so now both `r0` and `p` point to `x`). The `ldr` instruction on line 14 dereferences the pointer in `r0`. Specifically, it loads `r0` from the location `r0` points to (so it loads `x`). The `ldr` instruction on line 15 loads `r1` from `y`, and line 16 adds the value in `r0` (which is the value of `x`) and value in `r1` (which is the value of `y`) and leaves the sum in `r0`. The `ldr` on line 17 loads `r1` from `p` (recall the relative address of `p` is 2). So `r1` points to `x`. The `str` on line 18 stores `r0` (which contains the sum of `x` and `y`) in the location `r1` points to (which is `x`).

The `mov` instruction on line 20 has the effect of removing the local variable `y` from the stack. The `pop` instructions on line 40 and 41 restore `fp` and `lr` with the addresses they had on entry into `f`. Finally, the `ret` instruction on line 42 returns to startup code (by loading the `pc` register from `lr`).

Amux, Bmux, and Cmux Multiplexers

The `ld` instruction in the basic instruction set loads an operand from memory into the `ac` register. For example,

```
ld x
```

loads the operand at the label `x` into the `ac` register. Thus, after reading the operand from memory into the `mdr`, the microcode for the `ld` instruction transfer the operand in the `mdr` to the `ac` register. Here is the microcode for the `ld` instruction in the basic instruction set:

```
L0000:     ; ld ==========================================
           a.ir and b.m12 c.mar    ; move address in inst to mar
           rd                      ; read operand from memory
           a.mdr c.ac br@fetch     ; load ac from operand in mdr
```

The `ld` instruction in the optimal instruction set is more complex. It specifies not only the operand to be loaded but the register to load. For example,

```
ld r3, x
```

loads the operand at the label x into r3. The microcode for this instruction reads the memory operand into the mdr. But how does it get the contents of the mdr into r3? We can use

```
a.mdr c.r3 br@fetch     ; load r3 from operand in mdr
```

But this microinstruction is correct only if the ld instruction loads r3. What if the ld instruction loads any one of the other seven registers—for example,

```
ld r5, x
```

Note that the C field *of the microinstruction* above specifies the register to be loaded from the C bus. But we want the C field *of the machine instruction* to specify the register to be loaded. Then if the ld instruction is

```
ld r3, x
```

r3 is loaded (because the number of r3—011—is in the C field *of machine instruction*). But if the ld instruction is

```
ld r5, x
```

then r5 is loaded (because the number of r5—101—is in the C field of the *machine instruction*).

When we write microcode, sometimes we want a microinstruction that loads the register specified by the C field of the microinstruction in the mir. But at other times, as illustrated by the ld instruction in the optimal instruction set, we want a microinstruction that loads the register specified by the C field of the machine instruction in the ir. How do we provide the LCC this capability? Recall that the C decoder determines which register is loaded from the C bus. We simply add a multiplexer that drives the C decoder. The multiplexer selects either the C field in the microinstruction in the mir or the C field of the machine instruction in the ir. Let's call this multiplexer Cmux (the simulator program sim supports this extended model of the LCC):

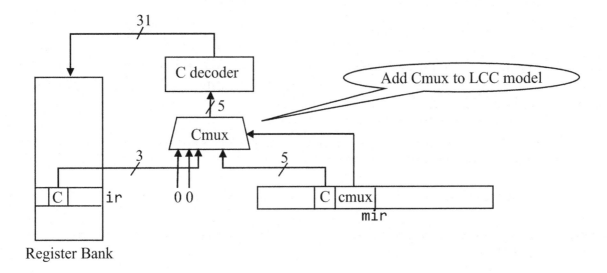

The Cmux multiplexer selects between the C field of the machine instruction and the C field in the microinstruction. The Cmux multiplexor has two five-line inputs. Because the C field in the machine instruction has only three bits, it is extended on the left with two 0's. The cmux field in the microinstruction (a one-bit field) determines which C field passes through to the C decoder. If cmux = 1, the C field in the machine instruction passes through; if cmux = 0, the C field in the microinstruction passes through.

In a symbolic microinstruction, if cmux is specified, then cmux = 1. Otherwise, cmux = 0. In the microcode for the basic and stack instruction sets, the register to load from the C bus is always specified by the C field in the microinstruction. Thus, we never need to specify cmux in the microcode for those instruction sets.

By selectively using the cmux field in a microinstruction, we can now microcode the ld instruction in the optimal instruction set:

```
L0010:      ; ld ==============================================
            a.ir sext b.m9 c.temp    ; get relative address
            a.temp add b.pc c.mar    ; convert to absolute address
            rd                       ; read operand from memory
            a.mdr cmux br@fetch      ; load destination reg from operand in mdr
```

The first microinstruction above extracts the pcoffset9 relative address from the instruction. The relative address is a signed number. Thus, we have to use the sext operation to extract it. The second microinstruction adds the relative address and the current contents of the pc to get the absolute address. The third microinstruction reads the main memory operand into the mdr. In the fourth microinstruction, no ALU operation is specified. Thus, the contents of the mdr (which are on the A bus) pass unchanged through the ALU to the C Bus. Because cmux is specified, the register specified by the machine instruction is loaded from the C bus.

The A decoder is configured with a multiplexer in the same way the C decoder is configured: The C decoder is driven by the Cmux multiplexer. Similarly, the A decoder is driven by the Amux multiplexer. The Amux multiplexer selects for output either the A field from the microinstruction or the A field in the machine instruction. The Cmux multiplexer is controlled by the cmux field in the microinstruction; the Amux multiplexer is controlled by the amux field in the microinstruction. The B decoder has a similar setup.

Let's take another look at the format of a microinstruction:

A	amux	B	bmux	C	cmux	ALU	u	rd	wr	cond	addr	
5	1	5	1	5	1	4	1	1	1	4	11	width

Note that it has the amux, bmux, and cmux fields that control the Amux, Bmux, and Cmux multiplexers. If these fields contain 1, then their corresponding multiplexers select the corresponding field in the machine instruction—not in the microinstruction. Armed with this knowledge, we can now write the microcode for the add instruction:

```
L0001:      ; add =============================================
            a.ir and b.bit5 !zer@add1      ; determine which type of add
add0        amux add bmux cmux u br@fetch  ; add two regs
add1        a.ir sext b.m5 c.temp          ; extract immed value
            amux add b.temp cmux u br@fetch ; add reg and immed value
```

The first microinstruction extracts bit 5 of the machine instruction. This bit determines the type of `add` instruction. If it is 1, a branch to `add1` occurs. If it is 0, then the microinstruction at `add0` is executed. Note that the microinstruction at `add0` specifies `amux`, `bmux`, and `cmux`. Thus, it is the machine instruction, not the A, B, and C fields in the microinstruction, that determines which registers are involved in the add operation. The microinstruction at `add1` is executed if the `add` instruction has an immediate operand. This microinstruction extracts the immediate operand and places it in the `temp` register. Then the next microinstruction performs the add operation. Note that it specifies `amux`. Thus, the operand to be added with the immediate value comes from the register specified by A field in the machine instruction field. Because `cmux` is specified, the destination register is determined by the C field in the machine instruction—not the C field in the microinstruction.

You may have noticed that u appears in two of the microinstructions above. We will explain its effect in the section below on user and system flag registers.

Complication with the St Instruction

Consider the following `st` instruction in the optimal instruction set:

```
st r1, x
```

It stores the contents of `r1` at the location corresponding to `x`. To do this, the microcode for the `st` instruction has to route the contents of `r1` down the A bus, through the ALU, to the `mdr`, in preparation for a memory write operation. The problem with doing this is that the register specified in a `st` instruction is *in its C field*, but we want it to drive the A bus. If we specify `amux` in a microinstruction, that causes the register in the A field—not the C field—of the machine instruction to drive the A bus.

The solution to this problem is quite simple: We shift the machine instruction in the `ir` three position to its right so that the register number specified in its C field ends up in the A field of the `ir`. Then a microinstruction with `amux` specified will drive the A bus with the register originally specified by the C field of the `ir`. Here is the microcode we need for the `st` instruction:

```
L0011:      ; st ===============================================
            a.ir sext b.m9 c.temp    ; get relative address
            a.temp add b.pc c.mar    ; convert to absolute address
            a.ir srl b.3 c.ir        ; shift reg number in C field to A field
            amux c.mdr               ; move reg contents to mdr
            wr br@fetch              ; perform write operation
```

User and System Flag Registers

Whenever the ALU performs an operation, it sets the flag registers according the results of the operation. For example, when an `add` instruction is executed, the ALU adds the two operands specified in the instruction. It sets `n` to 1 if the result is negative, `z` to 1 if the result is zero, `c` to 1 if a carry out of the leftmost position occurs, and `v` to 1 if signed overflow occurs. A conditional branch instruction can then test the flag registers and respond according to their settings. For example, consider the following sequence:

```
        add r0, r0, r1
        brp cat
```

The add instruction adds the contents of r0 and r1, and sets the flag registers according to the result. The brp instruction then branches if n = z. Because the result cannot simultaneously be both negative and zero, n and z cannot both be 1. Thus, if n = z, they both must be 0. But this implies the result of the addition is not negative (because n = 0) and not zero (but z = 0), and therefore is positive. Thus, the brp instruction branches if and only if the result of the addition is positive.

For the code sequence above to work correctly, the flag registers *must* remain unchanged from the time they are set by the add instruction to the time they are tested by the brp instruction. But before the brp instruction tests the flag registers, the CPU has to perform the fetch of the brp instruction, the incrementation of the pc register, and the decoding of the brp instruction. The incrementation of the pc register *sets the flag registers*. The decoding process uses sll instructions, *which also set the flag registers*.

So how can the flag registers remain unchanged, *as they must if our machine code is to work correctly*, from the time they are set by the add instruction to the time they are tested by the brp instruction? Here is the answer: They can remain unchanged because there are two sets of flag registers: a user set and a system set. When the add instruction performs the add operation, it sets the user flag registers. The incrementation of the pc register and the decoding process sets the system flag registers. Thus, when the brp instruction tests the flag registers (it tests the user set), they are unchanged from the time they were set by the add instruction.

The u field in a microinstruction determines which set of flag registers are used and set. If u = 1, the user set is used and set; if u = 0, the system set is used and set. For example, the actual add operation for the instruction

```
        add r0, r1, r2
```

is performed by the microinstruction at the label add0 in the following microcode:

```
L0001:      ; add ================================================
            a.ir and b.bit5 !zer@add1       ; determine which type of add
add0        amux add bumux cmux u br@fetch  ; add two regs
add1        a.ir sext b.m5 c.temp           ; extract immed value
            amux add b.temp cmux u br@fetch ; add reg and immed value
```

Because u is specified in the microinstruction at add0, its u field contains 1 (if u is omitted, the u field contains 0). Thus, the microinstruction sets the user flag registers. Now look at the microinstructions for the brp instruction that test the flag registers:

```
        u zer@fetch     ; go to next instruction if z = 1
        u neg@fetch     ; go to next instruction if n = 1
        a.temp add b.pc c.pc br@fetch ; branch to address in brp inst
```

The first two microinstructions test the flag registers. Because they have u, they test the user flag registers—the flag registers that were set by the add instruction. However, the microinstructions that increments the pc does not have u:

```
                ;==================================================
                ; Fetch machine instruction, increment pc
fetch           a.pc c.mar
                a.pc add b.1 c.pc rd     ← No u
```

Nor do the decoding instructions, for example,

```
L0:             a.ir sll b.1 c.dc neg@L01    ← No u
```

Thus, they set the system flag registers, leaving the user flag registers unchanged.

The trace that `sim` produces (recall that it is turned on if you enter "!" on the command line when you invoke `sim`) labels the system flag registers with "nzcv" and the user flag registers with "NZCV".

Epilog

Now that you have completed your study of the LCC from the transistor level to the assembly level, the next step is to study *C and C++ Under the Hood*, which covers the software side of the LCC—specifically, assemblers, linkers, compilers, interpreters, and of course, C and C++ under the hood.

Problems

1) Complete the microcode for the optimal instruction set in the file named `o.sm`. Test your microcode by assembling the program in `otest.a` by entering

   `optimal otest.a`

   Assemble your microcode by entering

   `micro o.sm`

   Then run the `otest` program using the simulator program by entering

   `sim otest.e`

   The `otest` program should display the numbers 1 to 25. Prefix the commands above with "./" on a Linux, Mac OS X, or Raspbian system. Hand in the ".1st" file created by the `micro` program and the ".log" file created by the `sim` program.

2) Draw the circuit that controls which set of flag registers is used.

3) What is the difference between a `ret` instruction and a `jmp r7` instruction?

4) In the IBM mainframe computers, the base register for relative addresses is specified at the beginning of each module (by a USING statement). The base register is loaded with the address of the *beginning* of the module. Thus, relative addresses are all non-negative. Compare this approach to the `pc`-relative

addressing used by the optimal instruction set. Discuss advantages and disadvantages of each approach.

5) The unused bits in the register version of the jsr instruction can be used to make the jsr instruction more versatile. Explain how.

6) The trap instructions work with r0. Suggest a modification that would allow the trap instructions to work with any of the eight registers r0 to r7 (so perhaps our optimal instruction set is not optimal).

7) Create a file ce0901.a that contains the optimal instruction set assembly code for the following C program. Assemble with the optimal program and run with sim. Hand in the ce0901.lst file created by the optimal program and the ce0901.log created by sim.

```
// ce0901.c
#include <stdio.h>
int y = 7, z;
int add10(int x)
{
    return x + 10;
}
int main()
{
    z = add10(y + 3);
    printf("%d\n", z);
    return 0;
}
```

8) Create a file ce0902.a that contains the optimal instruction set assembly code for the following C program. Assemble with the optimal program and run with sim. Hand in the ce0902.lst file created by the optimal program and the ce0902.log created by sim.

```
// ce0902.c
#include <stdio.h>
int a, y = 7;
void f(int x, int y, int z)
{
    int result;
    result = x + y - z;
    printf("%d\n", result);
}
int main()
{
    int b = 7, c;
    a = 1;
    c = a + b + y;
    f(a, b, c);
    return 0;
}
```

9) Create a file `ce0903.a` that contains the optimal instruction set assembly code for the following C program. Assemble with the `optimal` program and run with `sim`. Hand in the `ce0903.lst` file created by the optimal program and the `ce0903.log` created by `sim`.

```
// ce0903.c
#include <stdio.h>
void g(int x)
{
    printf("%d\n", x);
}
void f(int x)
{
    g(x - 2);
}
int main()
{
    f(5);
    return 0;
}
```

10) Create a file `ce0904.a` that contains the optimal instruction set assembly code for the following C program. Assemble with the `optimal` program and run with `sim`. Hand in the `ce0904.lst` file created by the optimal program and the `ce0904.log` created by `sim`.

```
// ce0904.c
#include <stdio.h>
int y = 7;
void f(int *p)
{
    *p = *p + 1;
}
int main()
{
    f(&y);
    printf("%d\n", y);
    return 0;
}
```

Appendix A: ASCII

Hex	Decimal	
20	32	<blank>
21	33	!
22	34	"
23	35	#
24	36	$
25	37	%
26	38	&
27	38	'
28	40	(
29	41	)
2A	42	*
2B	43	+
2C	44	,
2D	45	-
2E	46	.
2F	47	/
30	48	0
31	49	1
32	50	2
33	51	3
34	52	4
35	53	5
36	54	6
37	55	7
38	56	8
39	57	9
3A	58	:
3B	59	;
3C	60	<
3D	61	=
3E	62	>
3F	63	?

Hex	Decimal	
40	64	@
41	65	A
42	66	B
43	67	C
44	68	D
45	69	E
46	70	F
47	71	G
48	72	H
49	73	I
4A	74	J
4B	75	K
4C	76	L
4D	77	M
4E	78	N
4F	79	O
50	80	P
51	81	Q
52	82	R
53	83	S
54	84	T
55	85	U
56	86	V
57	87	W
58	88	X
59	89	Y
5A	90	Z
5B	91	[
5C	92	\
5D	93	]
5E	94	^
5F	95	_

Hex	Decimal	
60	96	`
61	97	a
62	98	b
63	99	c
64	100	d
65	101	e
66	102	f
67	103	g
69	104	h
69	105	i
6A	106	j
6B	107	k
6C	108	l
6D	109	m
6E	110	n
6F	111	o
70	112	p
71	113	q
72	114	r
73	114	s
74	116	t
75	117	u
76	118	v
77	119	w
78	120	x
79	121	y
7A	122	z
7B	123	{
7C	124	|
7D	125	}
7E	126	~

Important Control Characters

Hex	Decimal		Meaning
0A	10	\n	Line feed (i.e., newline)
0D	13	\r	Carriage return

Appendix B: Basic Instruction Set Summary

Opcode	Format		Description
0	ld	x	ac = mem[x];
1	st	x	mem[x] = ac;
2	add	x	ac = ac + mem[x];
3	sub	x	ac = ac - mem[x];
4	ldr	x	ac = mem[sp + x];
5	str	x	mem[sp + x] = ac;
6	addr	x	ac = ac + mem[sp + x];
7	subr	x	ac = ac - mem[sp + x];
8	ldi	x	ac = x;
9	asp	s	sp = sp + s;
A	call	x	mem[--sp] = pc; pc = x;
B	ret		pc = mem[sp++];
C	br	x	pc = x;
D	brz	x	if (ac == 0) pc = x;
E	brn	x	if (ac < 0) pc = x;
F	trap	y	see below

		Description
halt	or trap 0	Terminate program
nl	or trap 1	Output nl character
dout	or trap 2	Output number in ac as signed decimal
udout	or trap 3	Output number in ac as unsigned decimal
hout	or trap 4	Output number in ac in hex
aout	or trap 5	Output character in ac
sout	or trap 6	Output string pointed to by ac
din	or trap 7	Input decimal number into ac
hin	or trap 8	Output hex number into ac
ain	or trap 9	Input character into ac
sin	or trap 10	Input string to address in ac
bp	or trap 14	Breakpoint

x: bits 0 to 11 in machine instruction zero-extended to 16 bits
s: bits 0 to 11 in machine instruction sign-extended to 16 bits
y: bits 0 to 7 in machine instruction zero-extended to 16 bits
ac: accumulator register
pc: program counter register
sp: stack pointer register

Directives: .blkw, .fill, .start, .stringz

Appendix C: Stack Instruction Set Summary

Opcode	Format		Description
0	p	x	mem[--sp] = mem[x];
1	pi	x	mem[--sp] = x;
2	pr	s	mdr = mem[fp + s]; mem[--sp] = mdr;
3	cora	s	temp = fp + s; mem[--sp] = temp;
4	stav		temp = mem[sp++]; mem[mem[sp++]] = temp;
5000	dp		mem[sp] = mem[mem[sp]];
5001	esba		mem[--sp] = fp; fp = sp;
5002	reba		sp = fp; fp = mem[sp++];
5004	mhw		temp = mem[sp++]; mem[sp] = mem[sp] * temp;
5008	mmc		temp = mem[sp++]; mem[sp] = mem[sp] * temp;
6	asp	s	sp = sp + s;
7	add		temp = mem[sp++]; mem[sp] = mem[sp] + temp;;
8	sub		temp = mem[sp++]; mem[sp] = mem[sp] - temp];
9	call	x	mem[--sp] = pc; pc = x;
A	ret		pc = mem[sp++];
B	br	x	pc = x;
C	brn	x	if (mem[sp++] < 0) pc = x;
D	brz	x	if (mem[sp++] == 0) pc = x;
E	brp	x	if (mem[sp++] > 0) pc = x;
F	trap		see below

halt	or trap 0	Terminate program
nl	or trap 1	Output nl character
dout	or trap 2	Pop and display signed number in decimal
udout	or trap 3	Pop and display unsigned number in decimal
hout	or trap 4	Pop and display number in hex
aout	or trap 5	Pop and display ASCII character
sout	or trap 6	Pop address and display string at that address
din	or trap 7	Read decimal number from keyboard and push
hin	or trap 8	Read hex number from keyboard and push
ain	or trap 9	Read character from keyboard and push
sin	or trap 10	Pop address and read in string to that address
bp	or trap 14	Breakpoint

- x: bits 0 to 11 in machine instruction zero-extended to 16 bits
- s: bits 0 to 11 in machine instruction sign-extended to 16 bits
- y: bits 0 to 7 in machine instruction zero-extended to 16 bits
- sp: stack pointer
- fp: frame pointer register
- mdr: memory data register
- temp: temporary register

Directives: .blkw, .fill, .start, .stringz

Appendix D: Optimal Instruction Set Summary

Mnemonic	Format			Flags	Description
br--	0000	code	pcoffset9		on cond, pc = pc + pcoffset9
add	0001	C	A 000 B	nzcv	C = A + B
add	0001	C	A 1 imm5	nzcv	C = A + imm5
mov	0001	C	A 1 00000	nzcv	C = A
ld	0010	C	pcoffset9		C = mem[pc + pcoffset9)
st	0011	C	pcoffset9		mem[pc + pcoffset9] = sr
jsr	0100	1	pcoffset11		lr= pc; pc = [pc + pcoffset11]
jsrr	0100	000	A 000000		lr = pc; pc = baser
and	0101	C	A 000 B	nz	C = A & B
and	0101	C	A 1 imm5	nz	C = A & imm5
ldr	0110	C	baser offset6		C = mem[baser + offset6]
str	0111	C	baser offset6		mem[A + offset6] = sr
cmp	1000	000	A 000 B	nzcv	A − B (set flags)
cmp	1000	000	A 1 imm5	nzcv	A − imm5 (set flags)
not	1001	C	A 000000	nz	C = ~A
push	1010	C	000000001		mem[--sp] = C
pop	1010	C	000000010		C = mem[sp++];
srl	1010	C	000000100	nzc	C >> 1 (0 inserted)
sra	1010	C	000001000	nzc	C >> 1 (sign replicated)
sll	1010	C	000010000	nzc	C << 1 (0 inserted)
sub	1011	C	A 000 B	nzcv	C = A − B
sub	1011	C	A 1 imm5	nzcv	C = A − imm5
jmp	1100	000	A 000000		pc = A
ret	1100	000	111 000000		pc = lr
mvi	1101	C	imm9		C = imm9
mov	1101	C	imm9		C = imm9
lea	1110	C	pcoffset9		C = pc + pcoffset9
trap	1111	0000	trapvect8		see below

halt	or trap 0	Terminate execution
nl	or trap 1	Output newline
dout	or trap 2	Display signed number in r0 in dec
udout	or trap 3	Display unsigned number in r0 in dec
hout	or trap 4	Display number in r0 in hex
aout	or trap 5	Display ASCII character in r0
sout	or trap 6	Display string r0 points to
din	or trap 7	Read dec number from keyboard into r0
hin	or trap 8	Read hex number from keyboard into r0
ain	or trap 9	Read character from keyboard into r0
sin	or trap 10	Input string into buffer r0 points to
bp	or trap 14	Breakpoint

br mnemonic	code
brz or bre	000
brnz or brne	001
brp	010
brn	011
brlt	100
brgt	101
brc	110
br	111

Directives: .blkw, .fill, .start, .stringz

Appendix E: Microinstruction Format

A	amux	B	bmux	C	cmux	ALU	u	rd	wr	cond	addr	
5	1	5	1	5	1	4	1	1	1	4	11	width

Field

A   Specifies register that inputs to the A multiplexer

amux   Controls A multiplexer:
         amux = 0 then A field in `mir` drives A decoder
         amux = 1 then A field in `ir` drives A decoder

B   Specifies register that inputs to the B multiplexer

bmux   Controls B multiplexer:
         bmux = 0 then B field in `mir` drives B decoder
         bmux = 1 then B field in `ir` drives B decoder

C   Specifies register that inputs to the C multiplexer

cmux   Controls C multiplexer:
         cmux = 0 then C field in `mir` drives C decoder
         cmux = 1 then C field in `ir` drives C decoder

alu   Specifies ALU operation:

F_3	F_2	F_1	F_0		Mnemonic	Output	Flags Set
0	0	0	0	(0)	nop	left	
0	0	0	1	(1)	not	~left	nz
0	0	1	0	(2)	and	left & right	nz
0	0	1	1	(3)	sext	left sign ext, (rt = mask)	nz
0	1	0	0	(4)	add	left + right	nzcv
0	1	0	1	(5)	sub	left − right	nzcv
0	1	1	0	(6)	mult	left * right	nz
0	1	1	1	(7)	div	left / right	nz
1	0	0	0	(8)	rem	left % right	nz
1	0	0	1	(9)	or	left | right	nz
1	0	1	0	(10)	xor	left ^ right	nz
1	0	1	1	(11)	sll	left << right (logical)	nzc
1	1	0	0	(12)	srl	left >> right (logical)	nzc
1	1	0	1	(13)	sra	left >> right (arithmetic)	nzc
1	1	1	0	(14)	rol	left << right (rotate)	nzc
1	1	1	1	(15)	ror	left >> right (rotate)	nzc

u   User n, z, c, v flags used instead of system flags

rd   Initiates memory read from address in `mar`
wr   Initiates memory write of data in `mdr` to address in `mar`

cond   Specifies branch condition:

	Mnemonic	Branch if	Branch on
0	nobr		never
1	zer	z = 1	zero or equal
2	!zer	z = 0	not zero or not equal
3	neg	n = 1	negative
4	!neg	n = 0	not negative
5	cy or <	c = 1	less than (unsigned cmp)
6	!cy or >=	c = 0	grt or eq (unsigned cmp/overflow)
7	v	v = 1	signed overflow
8	pos	n = z	positive
9	lt	n != v	less than (signed cmp)
10	le	n != v or z = 1	less than or equal (signed cmp)
11	gt	n = v and z = 0	greater (signed cmp)
12	ge	n = v	greater than or equal (signed cmp)
13	<=	c = 1 or z = 1	less than or equal (unsigned cmp)
14	>	c = 0 and z = 0	greater than (unsigned cmp)
15	br		always

addr   branch-to address

Appendix F: Default Register Names

Reg Num	Name	Initial Contents	Function
0	r0 or ac		accumulator register
1	r1		
2	r2		
3	r3		
4	r4		
5	r5 or fp		frame pointer register
6	r6 or sp		stack pointer register
7	r7 or lr		link register
8	r8		
9	r9		
10	r10		
11	r11 or temp		
12	r12 or 3	0x0003	
13	r13 or 4	0x0004	
14	r14 or bit5	0x0020	
15	r15 or bit11	0x0800	
16	r16 or bit15	0x8000	
17	r17 or m3	0x0007	
18	r18 or m4	0x000f	
19	r19 or m5	0x001f	
20	r20 or m6	0x003f	
21	r21 or m8	0x00ff	
22	r22 or m9	0x01ff	
23	r23 or m11	0x07ff	
24	r24 or m12	0x0fff	
25	r25 or ir		machine instruction register
26	r26 or dc		decoding register
27	r27 or 1	0x0001	constant 1
28	r28 or pc		program counter register
29	r29 or mar		memory address register
30	r30 or mdr		memory data register
31	r31 or 0	0x0000	read-only register containing

Index

.a, 23
.blkw, 27
.fill, 23
.stringz, 27

A bus, 72
A decoder, 74
absolute address, 116
ac, 11
accumulator, 11
adder/subtractor circuit, 56, 61
address, 11
address-of operator, 96
ALU, 11, 65
ALU operation, 75
Amux, 127
AND circuit, 53
AND gate, 42
arithmetic/logic unit, 11
arithmetic-logic unit, 65
ASCII, 17
assembler, 25
assembly language, 23

B bus, 72
B decoder, 74
barrel shifter, 62
base-10, 1
base-16, 1
base-2, 1
basic, 25
basic instruction set, 13
bidirectional bus, 44
binary counter, 67
bit, 1
bit5, 79
bitwise operation, 53
Bmux, 127
Boolean function, 45
borrow technique, 59
branch instruction, 28
branch-control logic, 76
buffer, 27
byte, 1
byte-addressable, 12

C bus, 72
C decoder, 75

c flag, 60
c/b flag, 60
call, 31
called module, 31
caller, 31
calling module, 31
case insensitive, 24
central processing unit, 11
clear input, 51
clock, 49
clock sequencer, 69
clocked D latch, 49
Cmux, 126
command line, 16
command prompt, 16
comment, 16, 23, 24
conditional branch, 87
contents, 11
control unit, 11, 69
converting between binary and hex, 7
converting decimal to binary, 8
cora, 105
CPU, 11
cross-hatch, 53

data path, 72, 73
debugging microcode, 89
decimal, 1
decode, 12
decoder, 55
decoding the opcode, 84
dereference pointer, 123
dereferencing operator, 96
difference-detecting gate, 41
din, 21
directive, 23
divider circuit, 63
dout, 15
dp, 106

effective address, 116
enA, 65, 73
enB, 65, 74
esba, 100

fanout, 38
flop-flop, 50
fp, 100
frame pointer, 100

full adder, 47, 56

h2b, 16
half adder, 46
halt, 12, 15
hexadecimal, 1
high impedance, 43
high-Z, 43
hout, 21

immediate operand, 17
impedance, 43
infinite loop, 20
instruction register, 11
instruction set architecture, 116
interpreting machine code, 86
inverter, 40
ir, 11
ISA, 116

JK flip-flop, 50
jsr, 32

label, 23, 24
LCC, 11
ldi, 17, 26
ldr, 29
least significant bit, 3
load point, 12
log file, 16
loop, 12
lsb, 3

m8, 79
machine language, 11
mar, 63
mdr, 63
memory, 11
memory address bus, 73
memory address register, 63
memory data bus, 73
memory data register, 63
mhw, 107
micro, 87
micro-operation, 70
microprogram program counter, 67
microprogrammed computer, 11
microstore, 11, 64, 67
mir, 67, 76
mmc, 107
mnemonic, 23

MOS, 38
most significant bit, 3
mpc, 67, 76
msb, 3
multiplexer, 54
multiplier circuit, 63

n flag, 60
NAND gate, 43
nano, 15
negative edge, 50
newline character, 18
nl, 15
NMOS, 38
NOR gate, 42, 60
NOT circuit, 53
NOT gate, 40
notepad, 15
null character, 18
null-terminated string, 27
number of levels, 51

opcode, 14
opcode extension, 109
OR circuit, 53
OR gate, 42
overflow, 57, 120

parity, 71
pc, 11
pc-offset, 116
pc-relative addresses, 116
PMOS, 38
pointer, 96
pop operation, 29, 34
positional numbering system, 1
positive edge, 50
preset input, 51
program counter, 11
pseudo-instruction, 123
push operation, 29, 34

r31, 75
RAM, 63
random-access memory, 63
range of numbers, 6
rd, 73
read-only memory, 63
reba, 100
register, 11, 57, 65
register architecture, 99

relative instruction, 29
remainder circuit, 63
reset, 48
ret, 32
return address, 31
return character, 18
ROM, 63
rotate, 62

semiconductor, 38
sequencer circuit, 69
sequential circuit, 48
set, 49
SEXT circuit, 54
shift, 61
sign extension, 9
signed number, 5
signed overflow, 57
silicon, 38
sim, 16
sim.txt, 79
sin, 27, 28
source program, 25
sout, 18, 27
sp, 29
SR flip-flop, 50
SR latch, 48
stack, 29
stack architecture, 99
stack frame, 100
stack pointer, 11
startup code, 90
string, 17

subroutine, 31
symbolic microcode, 81
system flag registers, 129

Terminal, 16
transistor, 38
trap, 14, 20, 87
trap vector, 14
tri-state buffer, 43, 73
truth table, 41
two's complement, 4, 5
two-bus multiplexer, 55

unconditional branch, 20, 87
universal gate, 52
unsigned number, 5
unsigned overflow, 59
user flag registers, 129

v flag, 59
volatile memory, 64

whitespace, 18
word addressable, 12
word size, 11
wr, 73

XOR circuit, 53
XOR gate, 41

z flag, 60
zero extension, 9
zero-detecting gate, 42, 43

Made in the USA
Coppell, TX
12 January 2020